T0327543

Open RAN

Open RAN

The Definitive Guide

Edited by

Dr. Ian C. Wong
VIAVI Solutions
Austin, TX, USA

Dr. Aditya Chopra
Amazon Kuiper LLC
Austin, TX, USA

Dr. Sridhar Rajagopal
Mavenir
Richardson, TX, USA

Dr. Rittwik Jana
Google
New York, NY, USA

WILEY

Published by John Wiley & Sons, Inc., Hoboken, New Jersey.
Published simultaneously in Canada.

For general information on our other products and services or for technical support, please contact our Customer Care Department within the United States at (800) 762-2974, outside the United States at (317) 572-3993 or fax (317) 572-4002.

Wiley also publishes its books in a variety of electronic formats. Some content that appears in print may not be available in electronic formats. For more information about Wiley products, visit our web site at www.wiley.com.

Library of Congress Cataloging-in-Publication Data applied for:
Hardback ISBN: 9781119885993

Cover Design: Wiley
Cover Image: © Alexander56891/Shutterstock

Set in 9.5/12.5pt STIXTwoText by Straive, Pondicherry, India
Printed and bound by CPI Group (UK) Ltd, Croydon, CR0 4YY

C9781119885993_190923

Contents

List of Contributors

Sidd Chenumolu
DISH Network, Englewood
CO, USA

Prabhakar Chitrapu
SCF, St. Louis
MO, USA

Aditya Chopra
Amazon Kuiper, Redmond
WA, USA

Luis Manuel Contreras Murillo
Telefonica, Distrito Telefónica
MAD, Spain

Dhruv Gupta
AT&T, San Ramon
CA, USA

Rittwik Jana
Google, New York City
NY, USA

David Kinsey
AT&T, Seattle
WA, USA

Luis Lopes
Qualcomm, Boulder
CO, USA

Sridhar Rajagopal
Mavenir, Richardson
TX, USA

Diane Rinaldo
ORPC, Portland
ME, USA

Manish Singh
Dell, Round Rock
TX, USA

Rajarajan Sivaraj
Mavenir, Richardson
TX, USA

Padma Sudarsan
VMWare, Palo Alto
CA, USA

Reza Vaez-Ghaemi
VIAVI Solutions, Germantown
MD, USA

Ian Wong
VIAVI Solutions, Austin
TX, USA

Sameh M. Yamany
VIAVI Solutions, Boulder
CO, USA

Amy Zwarico
AT&T, Dallas
TX, USA

Foreword

Virtualization and cloud have unified computing and networking. This unification can be seen in all areas of wired communications. With Open RAN (O-RAN), wireless communications will be able to take advantage of the unification as well.

Wireless communications are overtaking wired communications. According to some estimates, traffic from wireless and mobile devices accounts for two-thirds of all IP traffic. However, the deployment of wireless communications is costly. The cost is mainly driven by RAN. For example, substantial increase in the number of cell sites and radio types with massive MIMO (multi-input multi-output) technology required by 5G drives RAN costs even higher. O-RAN attempts to solve this problem by disaggregating and splitting the RAN, supporting standardized interfaces that can interoperate with each other. This standardization drives multivendor radio, hardware, and software deployments with key functions implemented as virtual network functions (VNFs) and cloud-native network functions (CNFs) on vendor-neutral hardware and networks implemented in modular units.

Building O-RAN and ensuring its performance guarantees are huge undertakings. Compared to traditional RAN, installing and operating O-RAN is expected to be substantially more complex and time-consuming. The expense of this model may offset O-RAN cost savings as in virtualized wired networks. However, an open and virtualized model should drive more innovations, flexibility, and automation in wireless communications. For example, RAN Intelligent Controllers (RIC) with open APIs should provide a platform for the developer community to innovate RAN algorithms.

This book discusses various industry groups and standardization bodies participating in specifying O-RAN technologies; O-RAN architecture consisting of RIC, Service Management and Orchestration architecture, cloudification and virtualization, open transport, security, and open software; and typical O-RAN deployment scenarios. It is a valuable book in this area.

Mehmet Toy, PhD
Associate Verizon Fellow
Distinguished MEF Fellow
ITU-T SG13 (Future Networks) Vice Chair
IEEE Life-Time Senior Member

Preface

"Wouldn't it be great if we had a book on O-RAN?" It was a discussion I had with Sameh Yamany, VIAVI CTO and my manager and mentor, as I began my journey with VIAVI and in the world of open RAN and the O-RAN ALLIANCE. It was towards the end of 2020, where I first started as a contributor to the open fronthaul test specifications, and even after taking the helm as co-chair of the Test and Integration Focus Group. There's always something new that I learned as I delved deeper into the various topics encompassing the vast field of open RAN. As I then tried to on-board new employees in our company to contribute to the O-RAN ALLIANCE, or introduce open RAN to external audiences, I realized that a definitive guide written by active members and contributors to the specification body for Open RAN, i.e. the O-RAN ALLIANCE, would be a worthwhile contribution to the field. I first contacted my Ph.D. advisor, Prof. Brian Evans who suggested I contact a former graduate school buddy Aditya Chopra, in AT&T labs at the time, who was an active contributor to the open fronthaul specifications (WG4) in O-RAN. After a brief coffee conversation in the middle of the COVID pandemic, we sketched out a rough outline, and decided to bring on-board 2 more editors, Sridhar Rajagopal from Mavenir, and Rittwik Jana, from VMWare at the time, to embark on this journey. It was a long journey of recruiting expert authors, going through the editing process, and finally coming up with this book, which we hope would be useful guide to anybody new to the field of open RAN and wanting to know more about its ins and outs, but also to seasoned veterans in a specific area, wanting to understand more the other aspects of this vast field that is revolutionizing the telco industry as we speak.

Ian Wong, Editor, Open RAN,
The Definitive Guide

As one of the earliest O-RAN delegates, and lead delegate for AT&T in O-RAN Fronthaul Working Group, I was often approached by colleagues in both the technical and business organizations within AT&T requesting to learn more about Open RAN, its concepts and the role played by the myriad standards bodies within this ecosystem. It is not sufficient to just point to the documents or reports produced by the standards, as they are often unable to capture the nuances behind the decisions that are listed within them. So when Ian approached me about writing a book on Open RAN, we both quickly agreed on the unmet need of having a guide for newcomers into the world of Open RAN. Given the vastness of this ecosystem, we also knew it had to be an edited book with chapters authored by the experts in that particular slice of the Open RAN universe. With Ritwik and Sridhar, as well as the various chapter authors jumping on board, I am truly grateful for the wonderful learning experience I was afforded while working with such a brilliant group of accomplished leaders in their respective fields.

Aditya Chopra, Editor, Open RAN,
The Definitive Guide

At Mavenir, I have been working on Open RAN products since the inception of the O-RAN alliance, and contributing on open RAN specifications, eco-system development and solutions. I have always received and still continue to receive questions from our partners and customers on eco-system readiness, deployment aspects, security aspects, etc. This always required extra effort in customer education. While specifications are available, they are not in a form that can be easily consumed by someone wanting to understand open RAN. So, when Ian contacted me on the co-editing a book on open RAN, I felt it would be wonderful to provide all this information in one place for anyone investing in open RAN and putting together a team of expert contributors from various domains such as Open RAN product development, hardware, platform, test and measurement solutions, and operator deployments. It has been a pleasure working with my co-editors and contributors on this book and I look forward to this book being used as a definitive guide for open RAN.

Sridhar Rajagopal, Editor,
Open RAN - The Definitive guide

Ian and Aditya approached me to co-author this book on Open RAN early 2021 along with Sridhar. I was already heavily involved in the O-RAN standards organization and had gained a lot of experience driving proof of concepts at AT&T back then. It was a great opportunity to tell a story about a piece of technology that is moving so fast. A lot of hard work and passion goes into making a new piece of technology such as Open RAN real, from the crafting of standards specifications to open source implementations and eventually real world deployments. I thought that writing this book would be a nice way to capture some of those interesting tidbits to the engineers and practitioners of this evolving field.

We charted out a quick plan to capture the various aspects of Open RAN and invite experts in the field to contribute to this book. It was a long journey of getting all the chapter contributions incorporated into a draft and editing the overall narrative so that there is a flow. We hope that this "snapshot" of Open RAN technology will be a useful guide to all. I am very grateful and fortunate to work with so many talented people everyday in this field and it has truly been a humbling experience.

Rittwik Jana, Editor,
Open RAN - The Definitive guide

About the Authors

Dr. Ian C. Wong is presently the Director of RF and Wireless Architecture in the CTO office at VIAVI Solutions, where he is leading RF and wireless technology strategy, architectures, and standards. He is the co-chair of the Test and Integration Focus Group (TIFG) and the Next-Generation Research Platforms research stream in the O-RAN Alliance, and VIAVI's representative at the full member and steering groups of the NextG Alliance. From 2009 to Aug 2020, he was with NI (formerly National Instruments) where his last position was Section Manager of Wireless Systems R&D where he led the development of real-time end-to-end 5G wireless test and prototyping systems, and managed the company's 3GPP wireless standards and IP strategy. From 2007 to 2009, he was a systems research and standards engineer with Freescale Semiconductor where he represented Freescale in the 3GPP LTE standardization efforts. He is a senior member of the IEEE, was the Director of Industry Communities for IEEE Communications Society 2016–2019, and was the Industry Program Chair for IEEE Globecom 2014 in Austin.

Dr. Wong co-authored the Springer book "Resource Allocation for Multiuser Multicarrier Wireless Systems," numerous patents, standards contributions, and IEEE journal and conference publications. He was awarded the Texas Telecommunications Engineering Consortium Fellowship in 2003–2004, and the Wireless Networking and Communications Group student leadership award in 2007.

He received the MS and PhD degrees in electrical engineering from the University of Texas at Austin in 2004 and 2007, respectively, and a BS degree in electronics and communications engineering *(magna cum laude)* from the University of the Philippines in 2000.

Dr. Aditya Chopra is a Senior Communication Systems Engineer at Project Kuiper within Amazon, where he develops systems solutions and prototypes for Low-Earth Orbit communication networks. From 2017 to 2022, he was a Principal Member of Technical Staff at AT&T, where he performed communication systems research and prototyping of wireless cellular solutions. At AT&T, he was the lead representative at the Fronthaul Working Group (WG4) within the O-RAN Alliance. He was the lead representative and one of the key members developing fronthaul specifications within this group when it was part of xRAN. From 2012 to 2017, he was a systems engineer at National Instruments, developing high speed wireless test solutions. His research interests include wireless physical layer optimization and prototyping of advanced wireless communication systems.

Dr. Chopra received his PhD (2011) and M.S. (2008) degrees in electrical engineering from The University of Texas at Austin, Texas, USA and his B.Tech degree in electrical engineering from the Indian Institute of Technology, Delhi, India, in 2006.

Sridhar Rajagopal is currently SVP, Access Technologies at Mavenir Systems, where he leads cross-functional design and solutions for Mavenir's Open RAN products involving the RAN, radio and RIC platforms. Prior to this, he was one of the initial employees at Ranzure Networks, a cloud RAN start-up. He also had R&D roles in design, prototyping and standardization of 5G cellular and Wi-Fi systems at Samsung, in UWB technology at WiQuest communications and 3G/4G research at Nokia. He was an associate editor for the Journal of Signal Processing Systems (Springer) and has led leadership positions in standardization bodies such as IEEE and WiMedia. He was a co-recipient of the IEEE 2017 Marconi Prize Paper award for his research on mmWave systems. He has co-invented around 50 issued US patents. He received his M.S. and Ph.D. degrees from Rice University and is a senior member of IEEE.

Rittwik Jana is currently a Telco analytics and automation network engineer at Google. He was previously the chief architect of RAN intelligence at VMware and a Director/MTS of inventive science at AT&T Shannon Research Labs. He is an industry expert with 25 years of experience in numerous wireless technologies and standards. His work has focused primarily on building services for cellular network planning using AI/ML and Google cloud APIs, disaggregating and optimizing the RAN using Open RAN technologies, model driven control loop automation in O-RAN/ONAP and performance evaluation of next-generation mobile streaming apps such as AR/VR/360 video. He is the co-chair of the requirements group in Open RAN Software Community (OSC) and was the chair of WG5 FCC's Communications Security, Reliability, and Interoperability Council in 2021. He is a recipient of the 2016 AT&T Science and Technology medal and the 2017 IEEE Vehicular Technology society Jack Neubauer memorial paper award on full duplex wireless. He earned a Ph.D. in Telecommunications Engineering from the Australian National University, Canberra, Australia in 2000.

Definitions / Acronyms

3GPP	Third-generation Partnership Project (3gpp.org)
5GC	5G Core
5GS	5G System
5QI	5G QoS Class Identification
AAL	Acceleration abstraction layer
ACK	Acknowledgement
ACS	Adjacent Channel Selectivity
AF	Application Function
AI	Artificial intelligence
ARPU	Average revenue per user
AMF	Access and mobility management function
API	Application programming interface
AOI	Area Of Interest
AP	Application Protocol
A-PTS	Assisted-Partial Timing Support
ARP	Allocation Retention Priority
AS	Autonomous Systems
AWS	Amazon Web Services
BASTA	Base Station Antenna Standard
BBDEV	Baseband Development (Intel Library)
BBU	Baseband Unit
BGP	Border Gateway Protocol
BLER	Block Error Rate
BMCA	Best Master Clock Algorithm
BOM	Bill Of Material
BWP	Bandwidth Part
CaaS	Containers as a Service
CALEA	Commission on Accreditation for Law Enforcement Agencies
CAT	Cache Allocation Technology
CC	Component Carrier
CFR	Crest Factor Reduction
CI/CD	Continuous Integration/Continuous Deployment

CM	Configuration Management
CNCF	Cloud Native Computing Foundation
CNI	Container Network Interface
COTS	Commercial Off the Shelf
CP	Control Plane
CP	Cyclic Prefix
CPU	Central Processing Unit
CQRS	Command and Query Responsibility Segregation
CSAR	Cloud Service Archive
CSP	Communication Service Provider
CSR	Cell Site Router
CU	Centralized Unit
CUPS	Control User Plane Separation
CA	Carrier Aggregation
CN	Core network
COTS	Common or commercial off-the-shelf
CPRI	Common public radio interface
CQI	Channel Quality Index
cRAN	Cloud or centralized RAN
CTI	Cooperative Transport Interface
CUSM	Control, user, synchronization, and management plane, in relation to Open Fronthaul
dB	decibel
DC	Dual Connectivity
DC	Data Center
DHCP	Dynamic Host Configuration Protocol
DME	Data Management and Exposure
DMS	Deployment Management Service
DNS	Domain Name System
DPD	Digital Pre Distortion
DPDK	Data Plane Development Kit
DRAM	Dynamic RAM
DRB	Data Radio Bearer
DSCP	Differentiated Services Code Point
DU	Distributed Unit
DAST	Dynamic Application Security Testing
D-TLS	Datagram Transport Layer Security
DAPS	Dual Active Protocol Stack
DOCSIS	Data Over Cable Service Interface Specification
DL	Downlink
DM	Data Model
DUT	Device under test
EC2	Elastic Cloud Compute
ECR	Elastic Container Registry
EDC	Edge Data Center
EKS	Elastic Kubernetes Service

eMBB	enhanced Mobile Broadband
EMS	Element Management System
eMTC	enhanced Machine Type Communication
ETSI	European Telecommunications Standards Institute
EVC	Ethernet Virtual Circuits
EVM	Error Vector Magnitude
EARFCN	E-UTRA Absolute Radio Frequency Channel Number
eCPRI	Enhanced common public radio interface
eNB	evolved Node B, essentially the 4G base station
EN-DC	E-UTRAN New Radio Dual Connectivity
EPC	Enhanced packet core, core network for 4G
E2SM	E2 Service Model
E-UTRA	Evolved Universal Terrestrial Radio Access
E-UTRAN	Evolved Universal Terrestrial Radio Access Network
FCAPS	Fault, Configuration, Accounting, Performance, Security
FDD	Frequency Division Multiplexing
FEC	Forward Error Correction
FH	Fronthaul
FOCOM	Federated O-Cloud Orchestration and Management
FPGA	Field Programmable Gate Array
FQDN	Fully Qualified Domain Name
FAPI	Fronthaul Application Platform Interface
FHM	Fronthaul multiplexer
FHG or FHGW	Fronthaul gateway
FFT	Fast Fourier Transform
FM	Fault Management
GaN	Gallium Nitride
GCP	Google Cloud Platform
GKE	Google Kubernetes Engine
GM	Grand Master
GPS	Global Positioning System
GPU	Graphical Processing Unit
gNB	next generation Node B, essentially the 5G base station
gNB-CU	gNB Central Unit
gNB-DU	gNB Distributed Unit
GPDR	General Data Protection Regulation
HO	Handover
HT	Hyper Threading
HCP	Hyperscale cloud provider
HMTC	High Performance Machine Type Communication
I/O	Input/Output
IA	Intel Architecture
IaaS	Infrastructure as a Service
IE	Information Element

IFFT or iFFT	Inverse Fast Fourier Transform
IMS	Infrastructure Management Service or IP Multimedia Subsystem
IOC	Information Object Class
IEEE	Institute of Electrical and Electronics Engineers
IETF	Internet Engineering Task Force
ITU	International telecommunication union
IM	Information Model
IoT	Internet of Things
IOT	Interoperability Test
IP	Internet protocol
IPSec	Internet Protocol Security
K8S	Kubernetes
KPI	Key Performance Indicator
LCM	Life Cycle Management
LDC	Local Data Center
LEA	Law Enforcement Agency
LI	Legal Intercept
LLC	Last Level Cache
LLS	Lower Layer Split
LTE	Long Term Evolution
LAN	Local Area Networks
LSTM	Long Short Term Memory
MAC	Medium Access Control
MDT	Minimization of Drive Testing
ML	Machine Learning
MME	Mobile Management Entity
mMIMO	massive MIMO
MnS	Management Service
MOI	Managed Object Instance
MPLS	Multiprotocol Label Switching
MTBF	Mean Time Between Failures
MU-MIMO	Multi-User Multiple Input Multiple Output
MA	Managed Application
M2M	Machine-to-machine
MDP	Markov Decision Processes
ME	Managed Element
MF	Managed Function
MIMO	Multiple-input Multiple-output
MMS	Multimedia Message Service
mMTC	massive machine-type communication
MNO	Mobile network operator
NATS	Network Address Translation
NB-IoT	NarrowBand Internet of Things

NEBS	Network Equipment Building Systems
NETCONF	Network Configuration Protocol
NF	Network Functions or Noise Figure
NFO	Network Function Orchestrator
NGMN	Next Generation Mobile Network
NIC	Network Interface Card
NIS	Network Information Service
NMS	Network Management System
NUMA	Non-Uniform Memory Access
NAC	Network Access Control
NAS	Non-access-stratum
NBI	Northbound interface
Near-RT RIC	Near-Real-Time RAN Intelligent Controller
nFAPI	Networked Fronthaul Application Platform Interface
NFV	Network Function Virtualization
NG	Next Generation
NG-RAN	Next Generation RAN
NIST	National Institute of Standards and Technology
NTN	Non-terrestrial network
Non-RT RIC	Non-Real-Time RAN Intelligent Controller
NR	5G New Radio
NSA	non-stand-alone, in relation to 5G architecture
OAM	Operations, Administration and Maintenance
OFDM	Orthogonal Frequency Division Multiplexing
ONAP	Open Network Automation Platform
OOBE	Out of Band Emission
OS	Operating System
OTT	Over The Top
O-Cloud	O-RAN Cloud
O-CU-CP	O-RAN Central Unit Control Plane.
O-CU-UP	O-RAN Central Unit User Plane
O-DU	O-RAN Distributed Unit
O-eNB	O-RAN eNB
O-RAN	Open RAN, also short for O-RAN ALLIANCE *(o-ran.org)*
O-RU	O-RAN Radio Unit
OLT	Optical Line Terminal
ONU	Optical Network Unit
OFH or Open FH	Open Fronthaul
OpenRAN	Short for OpenRAN project group in Telecom Infra Project (TIP)
ONF	Open networking foundation *(opennetworking.org)*
OTIC	Open test and integration center
PA	Power Amplifier
PaaS	Platform as a Service
PAPR	Peak to Average Power Ratio
PCIe	Peripheral Component Interconnect Express

PCP	Public Cloud Provider
PDCP	Packet Data Convergence Protocol
PDSCH	Physical Data Shared Channel
PE	Provider Edge
P-GW	Packet Data Network Gateway
PHY	Physical Layer
PLMN	Public Land Mobile Network
PM	Performance Management
PMD	Poll Mode Driver
PNF	Physical Network Function
PRACH	Physical Random Access Channel
PRB	Physical Resource Block
PRTC	Primary Reference Time Clock
PSAP	Public Safety Answering Point
PSTN	Public Switched Telephone Network
PTP	Precision Time Protocol
PCI	Peripheral Component Interconnect
PDU	Protocol Data Unit
PKI	Public Key Infrastructure
PKIX	Public Key Infrastructure (X.509)
PON	Passive Optical Network
QoS	Quality of service
QoE	Quality of experience
QSFP	Quad Small Formfactor Pluggable
RAB	Radio Access Bearer
RAM	Random Access Memory
RAT	Radio Access Technology
RCEF	Radio Connection Establishment Failure
RDT	Resource Director Technology
RF	Radio Frequency
RFFE	RF Front End
RIC	RAN Intelligent Controller
RLC	Radio Link Layer
RLF	Radio Link Failure
RoE	Radio Over Ethernet
RRC	Radio Resource Control
RRH	Remote Radio Head
RRM	Radio Resource Management
RSRP	Reference Signal Received Power
RSRQ	Reference Signal Received Quality
RU	Radio Unit
Rx	Receive
RAN	Radio Access Network
RACH	Random Access Channel
rApp	Non-RT RIC Application

RF	Radio Frequency
RL	Reinforcement Learning
RIA RAN	Intelligence and Automation
ROMA RAN	Orchestration and lifecycle Management Automation
RRU	Remote Radio Unit
RT	Real Time
SCTP	Stream Control Transmission Protocol
SDAP	Service Data Adaptation Protocol
SDL	Specification and Description Language
SDU	Service Data Unit
S-GW	Serving Gateway
SLA	Service Level Agreement
SM	Service Model
SME	Service Management and Exposure
SMF	Session Management Function
SMO	Service and Management Orchestration
SMT	Simultaneous Multi Threading
SN	Sequence Number
SoC	System on Chip
SR	Segment Routing
SR-IOV	Single Root I/O Virtualization
SW	Software
SyncE	Synchronous Ethernet
SA	Stand-alone, in relation to 5G architecture
SAST	Static Application Security Testing
SBA	Service Based Architecture
SBOM	Software Bill of Materials
SCA	Software Composition Analysis
SCAS	Security Assurance Specifications
SCF	Small Cell Forum
SDN	Software Defined Networking
SDO	Standards development organization
SI	Systems integrator
SID	Segment identifier
SON	Self Organizing Network
SP	Service Provider
SSH	Secure Shell
SUT	System under test
T-BC	Telecom Boundary Clock
TDD	Time Division Duplex
TNL	Transport Network Layer
TSC	Telecom Slave Clock
TTI	Trasmission Time Interval
TTLNA	Tower Top Low Noise Amplifier
Tx	Transmit

TCO	Total Cost of Ownership
TDMA	Time-division multiple access
TIP	Telecom Infrastructure Project *(telecominfraproject.com)*
TLS	Transport Layer Security
TN	Transport Node
TNE	Transport network elements
TR	Technical Report
TRP	Transmission and Reception Point
TS	Technical Specifications
TU	Transport Unit
UE	User Equipment
UL	Uplink
UMTS	Universal mobile telecommunication system
UNI	Universal network interface
UP	User Plane
UPF	User plane function
uRLLC	ultra-reliable low-latency communications
VIP	Virtual Internet Protocol
VLAN	Virtual Local Area Network
VM	Virtual Machine
VPN	Virtual Private Network
VNF	Virtualized Network Function
vRAN	Virtualized RAN
WCDMA	Wideband code-division multiple access
WDM	Wavelength division multiplexing
WG	Working Group
xApp	Near-RT RIC Application
xNF	Any Network Function
YANG	Yet Another Next Generation (Data Modeling Language)
ZTA	Zero Trust Architecture

1

The Evolution of RAN

Sameh M. Yamany

VIAVI Solutions, Boulder, CO, USA

The last century marked the birth of the "Information Age," ushering humanity into an era where access to knowledge and information fueled a global digital economy that pretty much changed every aspect of our lives. At the center of this transformation was the role played by the Telecommunications industry. It all started with the creation of the first computer networks in the 1950s. Then the global and ready access to information exploded in the 1980s with the introduction of the World Wide Web. Next, the evolution of wireless communication technologies allowed access to information not only from universities, offices, and homes, but from anywhere in the world at any time of the day. Finally, while the first two generations of mobile services were mainly built to support person-to-person calls as well as innovative messaging services, it was the invention of smartphones at the turn of the twenty-first century, with instantaneous access to information, that led to the rise and proliferation of innovative apps and services that impacted the lives of billions of people around the world.

It is worth noting that the deployment of the fourth wireless generation (4G) of mobile technologies – often called the long-term evolution or LTE – was one of the fastest in telecommunications history, adding over 2.5 billion subscribers in less than 5 years compared to 10 years for the third wireless generation to reach the same number of subscribers. The fifth generation (5G) that started in 2019 is predicted to reach three billion subscribers by the end of 2025, making its adoption even faster than 4G.

The Radio Access Network, or RAN, has been a critical component to all mobile generations. This chapter describes how the RAN evolved with each generation and the factors leading to the need for a new Open Radio Access Network (O-RAN) architecture.

1.1 Introduction

According to the acclaimed book *Sapiens: A Brief History of Humankind* (Harari 2015), our kind – the modern humans – started to behave in a more intelligent and superior way some 70,000 years ago. One significant difference we had over the other archaic humans was our ability to imagine and envision a future state of our evolution and plan toward it. That unique feature transformed us from foragers to farmers and city dwellers. Then, around 500 years ago, the same quality ushered

Open RAN: The Definitive Guide, First Edition. Edited by Ian C. Wong, Aditya Chopra, Sridhar Rajagopal, and Rittwik Jana.

us into a scientific evolution, followed by an industrial revolution 250 years later that triggered our current information age era. Advanced communication was central to the industrial and information age revolutions. We can argue that smoke signals and drums were the first primitive modes of communication our kind utilized thousands of years ago. However, it was not until the early nineteenth century that the first electronic communication systems appeared.

As this book focuses on radio telecommunication, it is worth noting that the first use of radio signals in communications occurred in the 1920s. While the inventions of the telegraph and wired telephony had significant impacts on human life in the first half of the twentieth century, it was the people's appetite to be able to communicate with each other, on the move, from anywhere and at any time, that ushered us into the mobile telecommunication era. The first five generations of mobile services were developed between the 1980s and the 2020s (Figure 1.1).

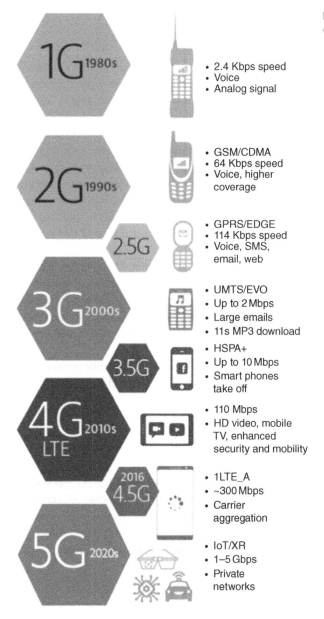

Figure 1.1 Five generations of mobile cellular networks.

1G 1980s
- 2.4 Kbps speed
- Voice
- Analog signal

2G 1990s
- GSM/CDMA
- 64 Kbps speed
- Voice, higher coverage

2.5G
- GPRS/EDGE
- 114 Kbps speed
- Voice, SMS, email, web

3G 2000s
- UMTS/EVO
- Up to 2 Mbps
- Large emails
- 11s MP3 download

3.5G
- HSPA+
- Up to 10 Mbps
- Smart phones take off

4G 2010s LTE
- 110 Mbps
- HD video, mobile TV, enhanced security and mobility

2016 4.5G
- 1LTE_A
- ~300 Mbps
- Carrier aggregation

5G 2020s
- IoT/XR
- 1–5 Gbps
- Private networks

Figure 1.2 Main components of a mobile network.

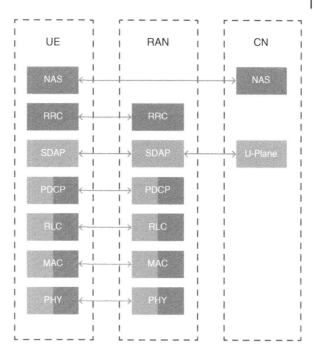

The RAN has been increasingly becoming the most valuable asset in mobile telecommunication systems and, over time, has shaped many of the innovations and acceleration of the information age revolution. A significant shift in RAN architecture was the move to packet-switched data to improve data traffic throughput in 3G. Another major re-architecture took place with the introduction of the LTE RAN. 5G RAN has been remodeled entirely to take advantage of software-defined networking (SDN) and the virtualization and cloudification of the network functions.

A simple mobile network comprises three main components, as shown in Figure 1.2. The core network (CN) ensures subscribers get access to the services they are entitled to. The CN also contains the critical functions of authentication, location, mobility, and performing the necessary switching tasks. The RAN connects the user equipment (UE) to other network parts through a radio interface.

The data and signaling communications between any mobile network architecture components are governed by a set of standards-defined protocols at every layer of interaction. A set of these protocols, as shown in Figure 1.3, determine how to set up and tear down signaling and communications between the mobile network elements. This set is mainly defined in the Control Plane or simply CP. Another set, called the User Plane or UP, enables the forwarding of data packets between the UE and the CN.

Several standards organizations develop and define mobile network protocols. All of them fall under the umbrella of the 3rd Generation Partnership Project (3GPP.org), which was established in 1998. The functional layers depicted in Figure 1.3 have all been specified and developed from Technical Specification (TS) documents within a 3GPP release. For example, the non-access-stratum (NAS) functional layer was specified by a particular 3GPP TS version to manage the setup and tear down of communications signaling between the UE and the CN and maintain these communication sessions as the UE moves throughout the network.

Another functional layer, the radio resource control (RRC), is responsible for the protocols between the UE and the Base Station. The RRC protocols' main functions include broadcasting system information, contacting the UE and paging them, and modifying or releasing any active

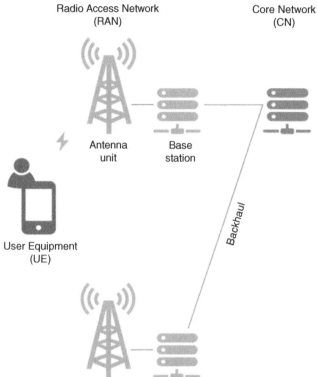

Radio Access Network
(RAN)

Core Network
(CN)

Figure 1.3 Layers of the protocol
stack of a 3GPP mobile network.

Antenna
unit

Base
station

User Equipment
(UE)

Backhaul

connections to the UE. It also takes care of the handover process between the different network-defined cells. Another primary function of the RRC protocol layer is to report the transmission signal strengths from the different cellular towers as received by the UE.

Other standard protocols for the RAN are the packet data convergence protocol (PDCP), responsible for the packet ciphering, data compression, and sequencing; the radio link control (RLC), responsible for error corrections and packets segmentation management; the media access control (MAC) performing the data multiplexing from different logical channels to be delivered to the air interface managed by the physical (PHY) functions.

1.2 RAN Architecture Evolution

To understand the history of advancements and innovations in mobile communication services, we must closely examine the drivers behind such explosive expansion over four decades going back to the 1980s. The primary force that ushered in the coming of the first generation (1G) mobile network was people's excitement and new expectations that they could communicate with each other from anywhere and at any time. Long gone was the need to find a wired-telephone post to call someone when your car broke down or you missed an important call from the office while shopping for groceries. Before cell phones became a reality, people used to call a place, not a person. Mobile communications changed that, and families and people found it easier to find each other at critical times, like being able to communicate emergencies or changes in kids pickup plans due to traffic jams.

The 1G voice services were purely analog and suffered from interference. As the number of users increased, it also lacked reliability and resiliency. The second generation mobile network, or 2G, moved from analog to digital, providing better quality and capacity. 2G also added a trendy feature, SMS, or Short Message Service, fashionably known as "texting."

Texting ushered humanity into a new culture of communication. We were able to send quick and efficient notes (and later with shorthands and emojis) to inform each other we would be late, set up a quick meeting, or announce a new addition to the family, all without the need to make a call. With MMS (Multimedia Message Service), we shared photos from anywhere in the world and celebrated together, almost in real-time, events thousands of miles away.

But we wanted more. We wanted to send emails on the go or access the internet from the malls. That is when improvement to 2G brought the 2.5G enhancements allowing the transfer of data packets over the mobile networks enabling subscribers to exchange email messages and browse the World Wide Web on the go.

As people's appetites for data access grew, the need for more capacity in mobile networks led to the third generation, or 3G, that supported much higher data rates than 2.5G, letting users exchange emails and large photos faster and with higher quality than before. Also, 3G and 3.5G witnessed the debut of video calls using a new type of mobile phone, smartphones.

But it was not until 4G, referred to as LTE, that we fully witnessed the beginning of the era of smartphones and mobile devices. LTE was also the first fully IP-based mobile network, thus enabling a plethora of innovative apps (short for mobile applications) encompassing every aspect of life, from music, food, shopping, TV, sports, games, social media, news, office and business, wellness, pet tracking, you name it.

While the first four generations of mobile networks were mainly focused on the users and subscribers, the explosive need for industrial automation, machine-to-machine communications, ultra-low latency remote applications, gaming, autonomous driving, and the emergence of AI-based IoT devices – all these new vertical industries required a completely new architecture for mobile networks. That was the start of the revolutionary 5G era.

1.2.1 The 2G RAN

The second generation (2G) mobile network design was essentially aimed at providing digital voice services and a moderate connection to data services. The 2G RAN architecture was labeled the Global System for Mobile communications or GSM, and it used TDMA (Time Division Multiple Access) for radio resource sharing, replacing the 1G analog systems.

The 2G RAN, shown in Figure 1.4, had two main components. The first was the BTS, or base transceiver station, which housed the radio resources. The deployment of several BTSs would be grouped within a particular geography and managed by one BSC or a base station controller. BSC supports many cellular functions within the BTS group, including handover scenarios, RF channel assignments, and the collection and maintenance of the different cell configuration parameters.

With the need to provide additional data access, the GSM, mainly circuit-switched-based, was improved to provide packet-based data connections. These middle-of-the-road improvements toward the third generation were called the GPRS, or General Packet Radio Services, sometimes labeled 2.5G. Still, the 2.5G RAN had a fundamental flaw; it was not designed from the ground up to support connection to all attached GPRS-capable phones even when they had very low data usage; hence, the third-generation RAN was needed with full-fledged packet-based data access architecture.

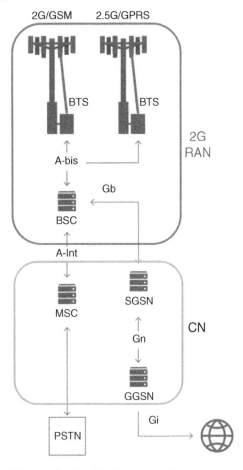

Figure 1.4 Simplified view of a 2G network.

1.2.2 The 3G RAN

The 3rd-generation mobile networks started in the late 1990s and early 2000s and were deployed initially in Japan, Europe, and the United States with the intent to co-exist and eventually replace the 2G/2.5G mobile infrastructure.

The 3G mobile RAN used a Wideband CDMA (WCDMA) radio physical layer and was technically labeled the UTRAN architecture, where UTRAN stood for universal mobile telecommunication system (UMTS) terrestrial RAN. The 3G RAN system, shown in Figure 1.5, had two new components; the first was the NodeB, replacing the BTS system in the 2G network, and the second was the radio network controller, or RNC, responsible for coordinating radio resources and mobility between several NodeBs.

The 3G deployment was a great success story in telecommunication history, and it fueled significant growth in subscriber numbers because it supported much higher data rates than the 2G system. Toward the mid-2000s, a new kind of cellular phone was being designed, the smartphone. However, the 3G network architecture did not foresee the demand from these new smartphones to be "always-on" and connected all the time to support new types of applications or apps running on these devices. These apps were characterized by low average data rates, with occasional high peak usage and low latency requirements.

An evolved 3G architecture was designed and implemented to address these shortcomings in the 3GPP release-five (Rel-5). The new RAN design used a High-Speed Downlink Packet Access or HSDPA for data access. The 3GPP Rel-6 defined another High-Speed Uplink Packet Access or HSUPA, and Rel-7 upgraded both these features and became known collectively as the HSPA+ enhancements, or more commonly, 3.5G, a precursor to the anticipated 4G 3GPP releases.

1.2.3 The 4G/LTE RAN

While most of the telecom service providers around the world were busy during most parts of the 2010s deploying and expanding their 3G and 3.5G networks, there was another threatening and competing technology based on the IEEE 802.16 standard on the rise – the WiMAX, short for Worldwide Interoperability for Microwave Access. WiMAX proponents claimed their technology already satisfied the major fourth-generation mobile network requirements defined by the International Telecommunication Union (ITU), in particular reaching speeds of up to 100 megabits per second, which was a leapfrog from the current 3.5G speeds. WiMAX also pushed to replace and dominate the last-mile broadband access for residential and enterprise customers. The threat

Figure 1.5 Simplified 3G/3.5G RAN architecture.

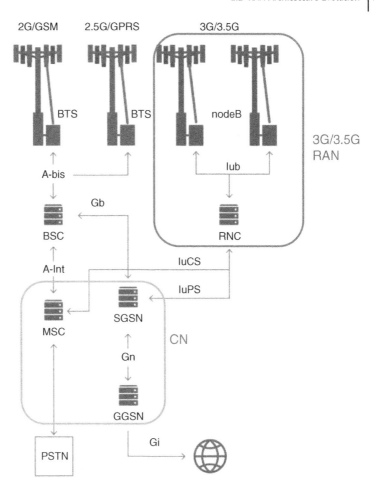

presented by WiMAX created a sense of urgency among the traditional telcos and led 3GPP to accelerate the release of the anticipated LTE standards. In addition to the WiMAX frenzy, the need to have a common standardized worldwide roaming motivated many telcos, particularly those that used CDMA in 3G, to migrate to the LTE standard much faster.

Figure 1.6 shows a simplified LTE RAN architecture. One of the significant differences from previous RAN designs was the elimination of the RNC. Instead, a new element was introduced, the evolved NodeB, or eNodeB, designed to connect to an IP-based network from the get-go. The eNodeB was smaller in size yet more efficient and high performing, benefiting from advances in microprocessor chips and smaller RAM components. LTE also expanded the use of a Remote Radio Head (RRH) connecting to the eNodeB with an optical fiber and using the Common Public Radio Interface (CPRI) standards for communication and managing the radio resources and the RF spectrum.

The initial deployment of LTE networks was the fastest in telecommunications history, yet the promised high-speed rates were not initially achieved. It was not until 3GPP Rel-10, famously known as LTE-Advanced, that the anticipated increased peak rates were achieved using a new technology called carrier aggregation, in which up to five carriers provided a 100 MHz total bandwidth. Another enhancement in Rel-10 was the use of Multi-Input Multi-Output (MIMO) antenna technology in which 8×8 MIMO was used for downlink and 4×4 MIMO for uplink.

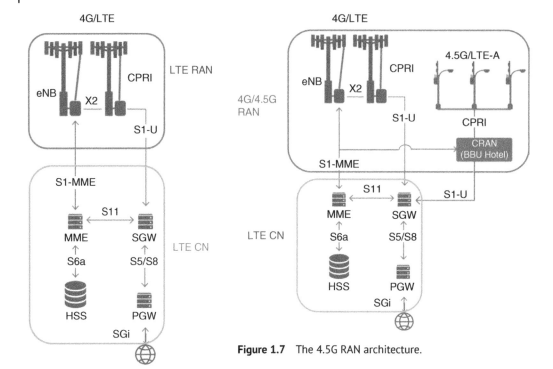

Figure 1.7 The 4.5G RAN architecture.

Figure 1.6 Simplified 4G/LTE RAN architecture.

With the fast expansion of smartphones and mobile device applications and the explosion of data traffic on the wireless network, it became apparent to the LTE network operators that the geographical distribution of traffic demand was not the same everywhere. As a result, coverage and capacity became significant concerns impacting customer experience and satisfaction. While macro cells provided connectivity at wider geographical ranges, the coverage performance at the edge of the cells suffered considerably. The idea of smaller cells was born to solve many of these issues. While small-cell antennas only cover smaller geographical areas, they provide better connectivity consistency, reliable coverage within their range, and more flexibility to combat interference and blind spots. This network densification led to an exponential increase in the number of RRHs, and it became cost prohibitive to attach an eNodeB to every RRH; hence came the introduction of a centralized pool of eNodeBs (sometimes called a BBU hotel) and the term centralized-RAN or CRAN was born. At the same time, 3GPP introduced releases 13 and 14, giving rise to the LTE Pro (also known as 4.5G). Figure 1.7 shows an example of the 4.5G RAN with small-cell and CRAN deployment.

By 2016, 4.5G networks provided increased data rates and higher bandwidth by enhancing carrier aggregation to support up to 32 simultaneous carriers, in addition to advancements in MIMO antennas from 16 to 64 elements. Another innovation in 4.5G was the introduction of two-dimensional beamforming technology. While beamforming was invented in the 1940s, it was not until the late 2010s when beamforming was introduced to wireless communication systems. In simple terms, beamforming technologies focus the radio signal transmission toward a target receiver, a cell phone in this case, rather than having the signal widely broadcasted in all directions. Such focus transmission can be performed horizontally in 2D or can be additionally narrowed down vertically in 3D.

1.2.4 The 5G RAN

The first two decades of the twenty-first century witnessed several advances in using artificial intelligence in many aspects of life. In particular, industrial automation and the rise of cloud service providers allowed enterprises and people to work and access their business applications and IT services remotely. They could scale up and scale down their computing and storage needs without too much upfront capital investment. Also, the threat from non-traditional communication providers and over-the-top players providing new innovative and free services led the traditional network operators to re-evaluate their strategy and design a new network architecture that offered differentiated services and features and justified the ROI on their infrastructure CAPEX investments.

In addition to the above, the global mobile ARPU (Average Revenue Per User) kept trending down. Yet, the new demand from industry verticals for the machine to machine (M2M) communications and everything connected to everything (e.g. automotive, drones, agriculture, utilities, airports, port authorities, and shipping) provided a new source of revenue besides subscribers' usage of the mobile services and the mobile M2M market was estimated to grow at a stunning double-digit CAGR percentage. However, these exciting new opportunities for the mobile network needed a new generational mobile network design to support essential features such as high throughputs, dynamic adaptability, network slicing, and ultra-reliable low latency (Grijpink et al. 2018). The new 5G network requirements – issued in 2015 by the ITU Radiocommunication Sector in their International Mobile Telecommuncations-2020 (IMT-2020) Standard – came to fulfill these promises (Figure 1.8).

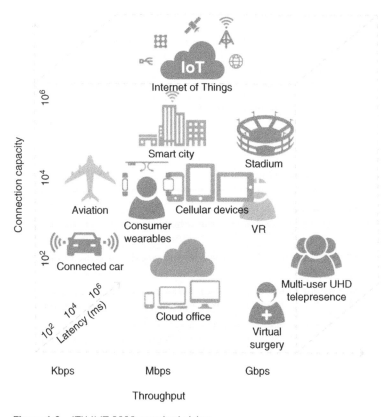

Figure 1.8 ITU IMT-2020-standard vision.

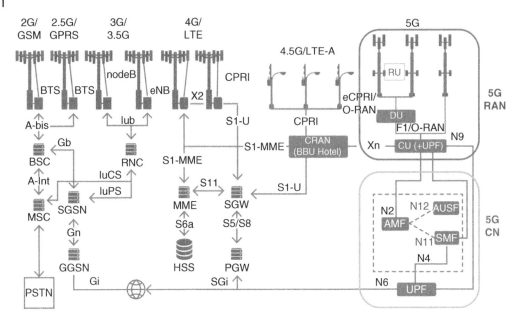

Figure 1.9 The evolution from 2G RAN to 5G RAN.

To deliver on the ITU IMT-2020 requirement, the 5G RAN architecture (shown in Figure 1.9) had to be disaggregated and provide flexibility and adaptability for different use cases that sometimes had contradicting and orthogonal demands.

For example, mMTC (massive Machine Type Communications) applications – e.g. traffic signs, parking management, shipping logistics, fleet mobility management, weather sensors, air and water quality sensors, charging and metering stations – required ultra-reliable low overhead to manage battery life for decades even if it meant higher latency and low bandwidth compromises. On the contrary, URLLC applications – e.g. remote surgery, emergency services, automotive-assisted and autonomous driving, and industrial automation – demanded high bandwidth, resiliency, and sub-millisecond latency, even if it meant bursts of higher energy consumption. Such contradicting requirements on the same RAN architecture meant the new 5G RAN must be disaggregated, highly flexible, and programmable. One example of such flexibility was the required ability to provide a User Plane Function (UPF) at different parts of the mobile network, hence disassociating it from the CN. In 5G, 3GPP defined a new part of the Next Generation NodeB (gNB replacing the eNodeB in LTE) as the Central Unit (CU) to house the UPF where needed. For more flexibility and programmability, the gNB can be split further with the CU providing support of higher protocol layers (such as PDCP, SDAP, and RRC) while a new physical entity, the Distributed Unit (DU), provided support for the lower protocol stacks (such as PHY, MAC, and RLC). Such an arrangement offered better scalability for different use cases, with one CU supporting multiple DUs and each DU connecting to multiple Radio Units or RUs. Each RU handled the radio signals sent to and from the 5G antennas, digitizing them for transmission over the packet networks.

One of the advantages of the new 5G RAN design was the cloudification aspects, where the DU software could be deployed on COTS (Common Off The Shelf) hardware with the option to run over private or public cloud infrastructure. In addition, the CU could also be deployed in the cloud, side by side with an edge computing platform, enabling critical applications to run much closer to the consumers at the edge of the network.

1.3   The Case for Open RAN

While the evolution to 5G opened up new revenue streams for the telecommunication industry beyond the traditional cellphone services, it presented a significant dilemma for the large, well-established telecom providers. As we explained before, the new 5G network was designed to provide flexibility, dynamic and real-time programmability, and fast adaptability to new vertical market requirements. That was the ticket for the traditional service providers to compete for head to head with the emerging actors in the communication sector, be it over-the-top players or the cloud-native providers who did not have any legacy infrastructure baggage to carry around and maintain backward compatibility. Yet, this transformation was not as simple as adopting a new technology design; it required cultural and operational shifts within the traditional telcos business models where the infrastructure deployed used to have an expected lifespan of 10–15 years and a payback time of multiple years. In particular, RAN elements were upgraded for mobile providers every 3–4 years and replaced every 10 years. The elements tended to have complex proprietary designs by the corresponding Network Equipment Manufacturer (NEM) and deployed on bespoke, high-performance, and purposely manufactured hardware. Typically, the hardware design was inflexible and required a costly and cumbersome forklift replacement with each wireless generation. With only a limited number of NEMs in the RAN domains, competing with each other for a more significant market share, most RAN elements used vendor-specific protocols. The interfaces between these RAN elements were competitively designed for optimal performance and only for that NEM-proprietary hardware, restricting any possible multi-vendor operability, as shown at the bottom layer of Figure 1.10. This vendor lock-in created a considerable cost on the traditional telcos, both from the CAPEX and OPEX expenditures, as well as time to market and the ability to provide innovative and differentiated services.

All of the above factors led the telcos to reassess the new network architecture and consider the shift toward an Open RAN system to break the NEM lock-in, stimulate market competitiveness, and spur innovation. The first attempt at openness resulted in the disaggregation of the hardware and software components of the RAN elements, as shown in the second layer from the bottom of Figure 1.10. In addition, some software components, especially in the CU element, were virtualized, coining the new term vRAN, or virtualized RAN. However, the interfaces at that stage were not fully open or standardized.

To achieve the desired full Open RAN, as shown in the top layer of Figure 1.10, several Open RAN groups (e.g. O-RAN Alliance, TIP) were set out to deliver well-defined specifications for a fully open RAN. These groups shared a common philosophy centered around designing and deploying multi-vendor, programmable, fully disaggregated, modular, and interoperable RAN network functions using open interfaces running on cloud-based virtual systems and leveraging COTS hardware. This allows operators to design and deploy mixed-vendor networks and network slices that is key to delivering several use cases in the same O-RAN infrastructure (Yamany 2021).

1.4   6G and the Road Ahead

The other significant development that will impact the future network beyond 5G is the ongoing 6G research in anticipation of a commercial launch in late 2030. The 6G research touches on four primary areas. The first is a new spectrum in the sub-terahertz and terahertz ranges. These new spectrum ranges will improve the efficiency, performance, and capacity of applications and services connectivity. The second area of 6G research is the native AI network design (Wu et al. 2021)

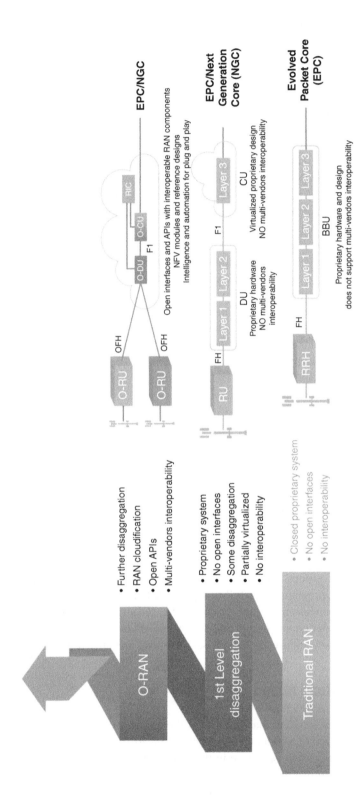

Figure 1.10 Evolution toward open RAN.

that will enable access to AI services and ML procedures in real-time from any network segment to any applications that require it. Such research will have a far-reaching impact on future network architecture and implementations. The third area of 6G interest and advancement will target the upcoming Non-Terrestrial Network (NTN) (Giordani and Zorzi 2020). NTNs are recognized as a critical element that will provide cost-effective and high-capacity connectivity and close the digital divide worldwide. The fourth area of 6G research concerns energy efficiency (Mahdi et al. 2021). While it is expected we will see an explosion in the number of devices and connectivity in 6G, new technologies are being researched to provide higher energy efficiencies and the possibilities of battery-free and sustainable energy sources.

1.5   Conclusion

The expectations of 5G and eventually 6G networks to support the new applications and verticals as we enter a new digital frontier with AI, ML, and automation will place enormous demand on the network infrastructure to deliver massive volumes of data over swatches of the spectrum to multitudes of users and devices at challenging latencies. The new RAN has been designed to meet these challenges. From the introduction of RAN function disaggregation and open interfaces in 5G to fully interoperable Open RAN architecture and ecosystem, to eventually a 6G RAN with native AI embedded, the role of RAN was and continues to be critical to the expansion of the telecommunications and digital services in every aspect of human life.

Bibliography

Giordani, M. and Zorzi, M. (2020). Non-terrestrial networks in the 6G era: challenges and opportunities. *IEEE Network* 35 (2): 244–251.

Grijpink, F., Ménard, A., Sigurdsson, H. et al. (2018). *The Road to 5G: The Inevitable Growth of Infrastructure Cost.* McKinsey & Company. https://www.mckinsey.com/industries/technology-media-and-telecommunications/our-insights/the-road-to-5g-the-inevitable-growth-of-infrastructure-cost (accessed 6 September 2022).

Harari, Y.N. (2015). *Sapiens: A Brief History of Humankind.* New York: Harper.

Mahdi, M.N., Ahmad, A.R., Qassim, Q.S. et al. (2021). From 5G to 6G technology: meets energy, internet-of-things and machine learning: a survey. *Applied Sciences* 11 (17): 8117. https://doi.org/10.3390/app11178117.

Wu, J., Li, R., An, X. et al. (2021). Toward native artificial intelligence in 6G networks: system design, architectures, and paradigms. *arXiv preprint arXiv:2103.02823.* https://arxiv.org/pdf/2103.02823.pdf.

Yamany, S. (2021). *O-RAN: an open ecosystem to power 5G applications.* VIAVI Solutions. https://www.viavisolutions.com/en-us/literature/o-ran-open-ecosystem-power-5g-applications-white-papers-books-en.pdf (accessed 6 September 2022)

2

Open RAN Overview

Rittwik Jana

Google, New York, NY, USA

2.1 Introduction

The digital society increasingly relies on wireless connectivity for a broad set of use cases, as more smartphones, vehicles, objects, and machines get connected to the internet. To support this, fifth generation (5G) cellular networks provide improved wireless communications and networking capabilities, enabling heterogeneous use cases such as:

- Evolved Mobile Broadband (eMBB) for high-definition multimedia use cases and immersive AR/VR streaming.
- Ultra-reliable low-latency communication (uRLLC) for industrial automation, intelligent transportation services, remote telesurgery, etc.
- Massive IoT use cases like smart grids, wearables, etc. requiring high coverage.

Beyond 5G cellular networks, such as 6G, focuses on enabling newer use cases like metaverse, telepresence, etc. that transcend boundaries of human imagination to deliver an unparalleled visual and perceptual experience to mobile users. The use case requirements and deployment scenarios keep changing with evolving radio access technologies. As a consequence, 5G and 6G cellular systems (and beyond) are expected to be highly complex systems deployed at a scale that is unforeseen in commercial networks so far.

The scale and complexity of 5G and 6G deployments, along with evolving use case requirements, have prompted research, development, and standardization efforts in newer RAN architectures, such as the Open RAN paradigm. Standardized by the O-RAN Alliance and based on the foundations of software-defined networking (SDN) and network function virtualization, the O-RAN architecture specifications are built on the following principles:

1) **Openness**: The interfaces between different functions or logical nodes in O-RAN architecture are open interfaces to achieve multi-vendor interoperability and co-existence across the functions.
2) **Virtualization**: The network function implementations in O-RAN architecture are migrated from vendor-proprietary hardware to commercial-off-the-shelf (COTS) cloud platforms running on whitebox hardware.

Open RAN: The Definitive Guide, First Edition. Edited by Ian C. Wong, Aditya Chopra, Sridhar Rajagopal, and Rittwik Jana.
© 2024 The Institute of Electrical and Electronics Engineers, Inc. Published 2024 by John Wiley & Sons, Inc.

3) **Intelligence**: RAN functions are open to radio resource management (RRM) by third-party optimization solutions deployed in a new centralized controller function, called the RAN intelligent controller (RIC), that performs closed-loop control of the RAN functions over open interfaces. These solutions leverage data-driven analytics and advanced AI and ML techniques to efficiently learn intricate inter-dependencies and complex cross-layer interactions between parameters across the layers of the RAN protocol stack toward optimizing RRM decisions, which cannot be captured by traditional RRM heuristics.

4) **Programmability**: The objective targets for optimization are programmatically configured and adapted using AI/ML-driven declarative policies, based on continuous monitoring of network and UE performance. Furthermore, the ML models for training and inference are updated using life cycle management to adapt to dynamics in the network, load, and traffic conditions.

Let us now elaborate on these fundamental pillars of Open RAN.

Openness: To bring about service agility and economies of scale, openness allows smaller vendors and operators to introduce customized services catered for each unique deployment environment. A multi-vendor solution (e.g. O-RU from vendor A and O-DU from vendor B) enables a more competitive supplier ecosystem that allows operators to mix and match the best-of-breed technologies. Finally, innovating on open-source software and hardware reference design enables quicker introduction of features in a more democratic manner. The figure below shows a multi-vendor Open RAN ecosystem. The benefit is that now with Open RAN an operator can change a network function (e.g., CU), and the rest of the ecosystem remains the same and performs as before (Figure 2.1).

Virtualization: Network functions like the IP Multimedia Subsystem (IMS), packet core (EPC), DUs, and CUs are deployed as virtual machines (VMs) or containers on COTS infrastructure. VMs provide an abstraction of the physical hardware that turns one server into many logical servers. The hypervisor allows multiple VMs to run on a single server. Virtualization of RAN

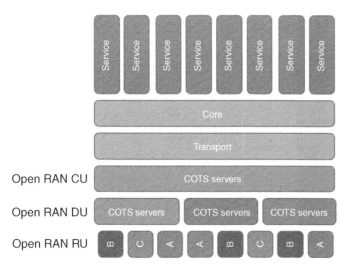

Figure 2.1 Multi-vendor Open RAN ecosystem.

(vRAN) is a new architecture whereby the baseband function is instantiated using a common set of resource pools made up of COTS servers situated in a centralized hub. vRAN provides outstanding value to operators by allowing the scaling up and down of serving capacity based on traffic demand, achieving a lower total cost of ownership (TCO) with a common hardware and software infrastructure. Nowadays, with a cloud-native approach, containerization technology provides further improvement for resource scaling at a micro-service level. Finally, with vRAN technology, network slicing services can be fully automated (instantiation, scaling, and continuous integration and continuous deployment, (CI/CD)) and reduce management complexities significantly in a multi-vendor ecosystem. Note that though RAN functions are virtualized on a COTS server, in vRAN, the interface between the BBU and RU is not an open interface. The key with Open RAN is that the interface between the BBU/DU and RU is an open interface, and as a result, any vendor's software can work on any open RU.

Intelligence: The near-RT RIC hosts intelligent applications as xApps. Applications like mobility management and interference management are available as apps on the controller, which enforces E2 network policies via a southbound interface toward the radios. The figure below shows how the near-RT RIC apps communicate with the radios and management system via E2 and A1, respectively (Figure 2.2).

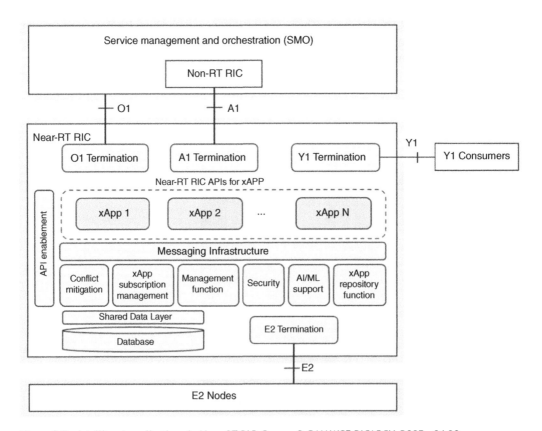

Figure 2.2 Intelligent applications in Near-RT RIC. *Source:* O-RAN.WG3.RICARCH-R003-v04.00.

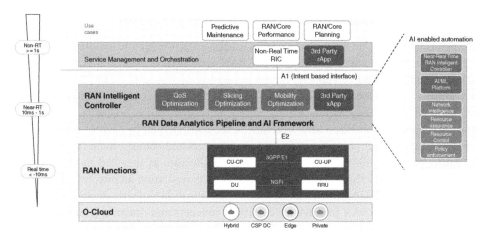

Figure 2.2a RAN Programmability

Programmability: Open RAN extends the programmability of RAN at a level that has not been possible before. By providing a consistent and standardized API, near-RT RIC apps can provide fine-grained precise control of radio capabilities (e.g. handover control, dual connectivity, carrier aggregation, etc.). Data can also be captured and explained in a structured manner that is useful to data scientists and machine learning (ML) model developers. Making the RAN more programmable using APIs allow the vast majority of the web developer community to now innovate and come up with interesting algorithms (Figure 2.2a).

2.1.1 What is Open RAN and Why is it Important?

Open RAN is the movement in wireless telecommunications to disaggregate hardware and software, to open interfaces, and reduce costs. With Open RAN and the "virtualization" it brings, operators are enabled to run software-based network functions on standard COTS servers. With non-proprietary, open interfaces, MNOs can use one supplier's radios with another's processors – something previously impossible. Networks need to be agile, flexible, and easily scalable. Hardware-centric, proprietary networks are not suitable for growing network demands and changing consumption patterns. Legacy networks often pose challenges for operators to expand and scale for coverage or capacity needs, and they are often difficult and costly to set up. Software at the center of the network delivers elasticity and scalability across all network components from access, to transport, and to core. A software-enabled approach provides unprecedented opportunities to bring down complexity, enhance operational efficiency and reduce costs.

2.1.2 How Does Open RAN Accelerate Innovation?

With more hardware and software vendors in the Open RAN ecosystem, the competition encourages innovation that is needed to meet the accelerated network demands of today. End users want faster speeds and lower latency to enable state-of-the-art applications that are enabled by 5G and beyond such as autonomous driving, internet of things (IoT), artificial intelligence (AI), streaming video, digital healthcare, and more.

2.1.3 What are the major challenges that Open RAN can help to address?

Two of the biggest challenges that networks face today are the cost to deploy and maintain networks due to vendor lock-in and network complexity. Open RAN can help unify different generations of access and core technologies while allowing MNOs to avoid vendor lock-in, extend their existing network investments, and make any generation (G) cellular deployments unified, simplified, and automated.

The following subsections provide a brief overview of the book chapter by chapter. Specifically, we discuss some of the salient points from each of the chapters and connect them together thematically. Chapter 3 focuses on the overall Open RAN architecture as specified in the O-RAN ALLIANCE, which comprises the Service Management and Orchestration architecture, RIC, and the identified functional splits and important use cases. Cloudification and virtualization concepts are introduced in Chapter 4. RAN intelligence controller concepts and design are introduced in detail in Chapter 5 highlighting a few important use cases. The next few chapters discuss the open fronthaul, open transport, security, and open-source software. Finally, we end with a detailed discussion of typical Open RAN deployment scenarios and the level of testing and specifications that are available to bring Open RAN to maturity and other industry organizations that enable the Open RAN ecosystem. The following subsections summarize the contents of the subsequent chapters in the book.

2.2 Open RAN Architecture

Chapter 3 provides a detailed view of the architectural components of Open RAN. O-RAN architecture integrates a modular base station software stack on off-the-shelf hardware, which allows baseband and radio unit components from discrete suppliers to operate seamlessly together. Open RAN disaggregates the network into the Radio Unit (RU), Distributed Unit (DU), and Centralized Unit (CU). The RU is where the radio frequency signals are transmitted, received, amplified, and digitized. The DU and CU are the computational parts of the base station, sending digital radio signals into the network. With Open RAN architecture, the interfaces or protocols between these building blocks and the software are open and non-proprietary. O-RAN leverages the functions and interfaces introduced by 3GPP, and the intent is not to duplicate what is already specified there. The main areas of extension are the open FH interface between the O-RU and O-DU, enhancing orchestration and management interfaces with O1/O2 definition and introduction of the near-RT RIC and the non-RT RIC with the R1, A1, and E2 interfaces for optimizing the network.

In order to optimize the RRM decisions associated with these functions, O-RAN introduces two RIC components, which include the near-RT RIC and the non-RT RIC, and additional open interfaces through which crucial RAN information is exchanged with the RIC components. The near-RT RIC is closer to the edge of the RAN and is responsible for controlling RRM decisions at a per-UE level in near-RT granularities using extensible third-party applications (xApps). The non-RT RIC is deployed in a centralized cloud within an SMO, sets high-level declarative RRM policies, and recommends configurations of parameters at UE-level, UE group-level, slice-level, cell-level, etc. for the near-RT RIC and the SMO, respectively, through third-party non-RT RIC apps (called rApps).

The SMO uses these recommendations for FCAPS operations to optimize the Operations, Administration and Maintenance (OAM) functionalities in the underlying network elements, while the near-RT RIC uses these policies for fine-grained RRM in the underlying RAN functions. The SMO can also be further used for the orchestration of computational and storage resources in the cloud. Utilizing the infrastructure and deployment management functions, the SMO will orchestrate the life cycle management of the application workloads by appropriately associating the compute and storage resources in the cloud.

The SMO includes the non-RT RIC, and the O-RAN architecture assigns some specific SMO functions (SMOFs) as being anchored inside or outside the non-RT RIC Framework. There are several SMOFs that can be implemented in different ways by different solution providers. There is a task force in O-RAN on how to improve O-RAN SMO architecture to support multi-vendor inter-operability, and it looks at the integration of existing standardized functions with new ones defined in O-RAN. There is a strong need for O-RAN to identify and decouple SMOFs and the related SMO services (SMOSs) within the SMO for the following reasons:

a) Operators could have legacy systems covering some SMOFs which need to integrate with other newer SMOFs.
b) Operators could integrate SMO into their existing network, management, orchestration, and RAN control elements by consuming the services provided by SMO.
c) Deployment and scaling considerations could indicate a natural separation of functions into SMOFs such that there is an M : N mapping of the number of different SMOFs in an operator's network.
d) SMOFs could be supplied by different vendors.
e) RAN-specific and generic functions are likely to have different development cycles.
f) The operational requirements, deployment, scaling, and life cycle management of different components could be different.

Using Open interfaces and Open APIs, the O-RAN architected components communicate with each other, and this helps foster a competitive and innovative ecosystem.

2.3 Open RAN Cloudification and Virtualization

Service providers want to improve RAN flexibility and deployment velocity, while at the same time reduce the capital and operating costs through the adoption of cloud architectures with flexibility to deploy/lease or reuse hardware. Cloud deployment strategy is generally to "Build Once, Deploy and Manage on Any Cloud" and more and more customers are choosing multi-cloud cloud options and associated platform services.

A key principle is the decoupling of RAN hardware and software and exposing the hardware components as logical units to the SMO. Utilizing the infrastructure management function and deployment management functions, the SMO will orchestrate the life cycle management of the application workloads by appropriately associating the compute resources. The O-RAN-defined O2 and the AALI interfaces enable applications to utilize the underlying infrastructure and plat-form services minimizing custom implementation.

Chapter 4 will cover:

- Key principles identifying the standards and open-source enablers for cloud-based deployments.
- Abstracting applications from the infrastructure to allow for application portability across a multi-cloud environment where regional and edge/far edge cloud compute resources (infra-structure and platform) can vary substantially.
- Flexible instantiation and life cycle management through orchestration automation to reduce deployment and ongoing maintenance costs.
- Challenges and opportunities with disaggregation of vertical and horizontal stack.

2.4 RAN Intelligence

One of the pillars of Open RAN is to provide developers an easy and open way to introduce intelligent behavior in the RAN. Specifically, we offer a detailed view of near-RT RIC and non-RT RIC and their APIs that allows the developer community to innovate RAN algorithms swiftly. Traditional radio vendors have predominantly owned innovation in the RAN domain. Accelerating innovation in the RAN domain is much needed for 5G to realize new use cases such as autonomous driving, IoT, AI, low latency streaming video and digital healthcare just to name a few. With Open RAN, optimization and automation of RAN algorithms can now be opened up to the worldwide developer community allowing the operators to choose the best-of-breed algorithm for deployment. Mobile networks across the globe are poised to benefit from three key technology areas: advances in SDN, the evolution of cloud-based radio access networks (RAN), and rapid progress in the fields of AI and ML. These technologies offer tremendous new opportunities for operators, not only by enabling innovative services and new revenue streams, but by also achieving significant cost reduction by applying intelligent automation in all aspects of a mobile network. Intelligence across the network, and within individual network functions, is expected to drive innovative solutions for network planning, engineering, and operation.

There are three key enablers required to bring such an intelligent and autonomous network to fruition: the ability to monitor the network with the right data set available in a timely manner, a rich ecosystem of AI/ML-driven applications to optimize and heal the network, and lastly, the ability to control and guide the network based on operator's business requirements. O-RAN-defined interfaces (O1, A1, E2) and controller platforms (non-RT and near-RT RIC) aim to enable this vision.

Chapter 5 will cover the following areas, namely:

- ML-driven intelligence and analytics for non-RT RIC
- ML-driven intelligence and analytics for near-RT RIC
- Current challenges and opportunities in building intelligent networks
- Background on ML life cycle management
- E2 service models and ML algorithms for near-RT RIC

2.5 Fronthaul Interface and Open Transport

Chapter 6 provides an overview on the O-RAN Open Fronthaul Interface. We describe existing methods of fronthaul operation with CPRI and eCPRI-based messaging and proprietary formats. Then we provide an overview of 3GPP studied physical layer split options and describe the advantages and disadvantages of different split options. This leads directly into the considerations behind O-RAN 7.2 split, and we describe this split option in detail. We show how this split operates in different types of deployment such as mmWave with analog beamforming, or massive MIMO with digital beamforming. We also describe the management plane interface and the static, semi-static, and dynamic management considerations, and fronthaul management topologies supported by this interface.

Following this, Chapter 7 provides a comprehensive overview of open transport (for fronthaul, midhaul, backhaul, and synchronization) requirements followed by a series of solutions that address these requirements. Specifically, we provide WDM, packet-switched, and synchronization solutions. Specifically in WG9: Open X-haul Transport Work Group, the focus is on the transport domain, consisting of transport equipment, physical media, and control/management protocols associated with the transport network.

2.6 Securing Open RAN

How is security ensured in Open RAN networks?

By complying with open and standard interfaces from organizations such as the O-RAN Alliance and implementing security testing with TIP member organizations, MNOs can ensure that their networks are interoperable and secure. Specifically, in this chapter, we discuss the zero-trust principles, threats to open RAN, and strategies employed to protect Open RAN. From its inception, security has been part of the O-RAN design. The security controls in the architecture are driven by zero-trust architecture principles, standards-based security technologies, and 3GPP security requirements. This simplifies security implementation for vendors, provides a clear migration path as the technologies evolve, and allows operators to more easily deploy security controls.

O-RAN security mandates authentication with digital certificates on all O-RAN-defined interfaces, plus encryption on all interfaces except for the Open Fronthaul S-plane interface, where encryption remains technically impractical. Authentication and encryption are basic controls in any zero-trust architecture. Layered onto these basic security requirements are controls that help protect the supply chain: secure development practices when developing O-RAN components and the inclusion of an NTIA-compliant software bill of materials with the component. The RICs enable operators to enhance RAN security with xApps and rApps that perform AI and ML event collection and analysis within the RAN.

Chapter 8 begins with a review of the practical threats against an O-RAN deployment and a review of zero-trust architecture principles. We then cover in detail the controls delivered with each O-RAN component. This includes a discussion of the standards-based protocols, ciphers, and other configurations used in the controls. We next discuss the supply chain security afforded by O-RAN. We end by discussing how an operator can use xApps and rApps to enhance attack detection and mitigation.

2.7 Open Source Software

The shift to virtualized RAN and open RAN is creating new opportunities for carriers to shift much of their RAN hardware and software to server-based platforms and solutions based on open-source software [Windrvr] (2021). The O-RAN Alliance and other industry groups have both developed and adopted open-source solutions for significant parts of the open RAN functionality (Figure 2.3) (Heavy Reading 2021). These solutions are already seeing early deployments, and much of the industry is expecting to use software infrastructure based on open-source solutions to run RAN applications. Chapter 9 provides an overview of the various open-source software efforts for realizing Open RAN.

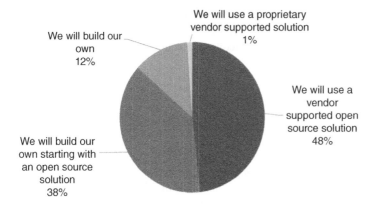

Figure 2.3 Accelerating open RAN platforms operator survey. *Source:* Adapted from [Heavy Reading, May 2021].

2.8 RAN Automation and Deployment with CI/CD

How can Open RAN enable cost savings?

Most of the CAPEX required to build a wireless network is related to the RAN segment, reaching as high as 80% of the total network cost. Reductions in the RAN equipment cost will significantly help the bottom line of wireless operators as they struggle to cope with the challenges of ever-increasing mobile traffic and flat revenues. Analysts' projections of a 5G macro-cell deployment with open architecture show pricing falling by 50% until 2022, whereas it will only fall by 30% if it is built traditionally with proprietary hardware and software. This 20% difference equates to hundreds of millions of dollars in the overall Total Cost of Ownership (TCO) and will help mobile operators extend investments and become more profitable. Open RAN is the key to cost-effectively deploying next-generation mobile network infrastructure. We discuss the various automation and deployment perspectives of Open RAN that are necessary to enable cost savings.

Chapter 10 provides insights and learnings from the world's first O-RAN 5G deployment in a true hybrid cloud environment. It will describe the approach, common and not-so-obvious pitfalls, benefits, and organizational processes. The chapter covers O-RAN and 3GPP interoperability with Devices and Core, how an operator can embrace quick releases, fail fast with test automation framework, and agile release management. It will include details on infrastructure, platform selection process, the importance of Zero Touch Provisioning for site turn-up process, interoperability between RU and DU, PTP, LCM, hardware dimensioning, FCAPS, CICD, network orchestration, and analytics. It will also touch on practical considerations to deploy O-RAN in Centralized vs. Distributed RAN and system reliability with COTS. The chapter will describe how a system can be made high availability with COTS hardware using cloud technologies, orchestration, and assurance. It will describe at high level, implications and considerations to implement near-RT RIC and non-RT RIC. The chapter concludes with some tips on how to make the network future proof.

2.9 Open RAN Testing

Multi-vendor interoperability is one of the most critical elements for the success of the Open RAN ecosystem. In order to ensure this, robust and comprehensive testing and integration are absolutely necessary and are in fact much more important for Open RAN compared to traditional single-vendor RAN. Chapter 11 outlines the key challenges and potential solutions to the testing and integration of multi-vendor Open RAN equipment. This includes the efforts within the O-RAN ALLIANCE to develop robust test specifications, the creation of Open Test and Integration Centers (OTICs), the processes being developed around certification and badging of equipment, and the role that O-RAN Global PlugFests play. We also outline the key role of 3GPP test specifications and how they complement and interact with O-RAN test specifications. Finally, we outline key test methodologies that are needed for the full life cycle, from lab-based testing, to early field tests and deployment, to full turn-up and operation, and finally to maintenance and operator assurance needs.

2.10 Industry Organizations

Each successive generation of cellular standards increases the diversity of deployments and use cases that the cellular network must support. This causes a corresponding increase in the size of the device and interface ecosystem that makes up a RAN. Consequently, a goal such as Open RAN is difficult to achieve by a single standards defining organization, and a variety of such organizations must work in tandem to realize the goals of Open RAN. We highlight some of these in Chapter 12 and provide a brief introduction for each organization.

2.11 Conclusion

We have provided comprehensive coverage of Open RAN technologies that will benefit practitioners of the new disaggregated RAN world. This book dives into the necessary specification details (in terms of architecture, protocols, APIs, and interoperability testing scenarios) and also real-life deployment examples of the world's first O-RAN 5G deployment in a true hybrid cloud environment. In addition, we showcase the various open-source activities that are thriving in the community to make Open RAN a fertile landscape for innovation.

Bibliography

Heavy Reading (2021). *Accelerating Open RAN Platforms Operator Survey*, May 2021.
[Windrvr] (2021). *WindRvr Blog*. https://blogs.windriver.com/wind_river_blog/2021/06/open-source-software-expected-to-drive-open-ran-deployments.

3

O-RAN Architecture Overview

Rajarajan Sivaraj and Sridhar Rajagopal

Mavenir, Richardson, TX, USA

3.1 Introduction

Standardized by the Open Radio Access Network (O-RAN) Alliance and based on the foundations of software-defined networking and network function (NF) virtualization, the O-RAN architecture specifications are built on the following principles:

1. **Openness**: The interfaces between different functions or logical nodes in O-RAN architecture are open interfaces in order to achieve multi-vendor interoperability and coexistence across the functions.
2. **Virtualization**: The NF implementations in O-RAN architecture are migrated from vendor-proprietary hardware to commercial-off-the-shelf (COTS) cloud platforms running on white-box hardware.
3. **Intelligence**: RAN functions are open to radio resource management (RRM) by third-party optimization solutions deployed in a new centralized controller function, called the RAN Intelligent Controller (RIC), that performs closed-loop control of the RAN functions over open interfaces. These solutions leverage data-driven analytics and advanced artificial intelligence (AI) and machine learning (ML) techniques to efficiently learn intricate interdependencies and complex cross-layer interactions between parameters across the layers of the RAN protocol stack toward optimizing RRM decisions, which cannot be captured by traditional RRM heuristics.
4. **Programmability**: The objective targets for optimization are programmatically configured and adapted using AI/ML-driven declarative policies, based on continuous monitoring of network and user equipment (UE) performance. Furthermore, the ML models for training and inference are updated using life cycle management to adapt to dynamics in the network, load, and traffic conditions.

3.1.1 General Description of O-RAN Functions

This section provides a general description of the O-RAN functions. An informal view of the different O-RAN functions, RIC/SMO functions along with xApps and rApps, along with their relationship cardinality, is provided in Figure 3.1.

Open RAN: The Definitive Guide, First Edition. Edited by Ian C. Wong, Aditya Chopra, Sridhar Rajagopal, and Rittwik Jana.

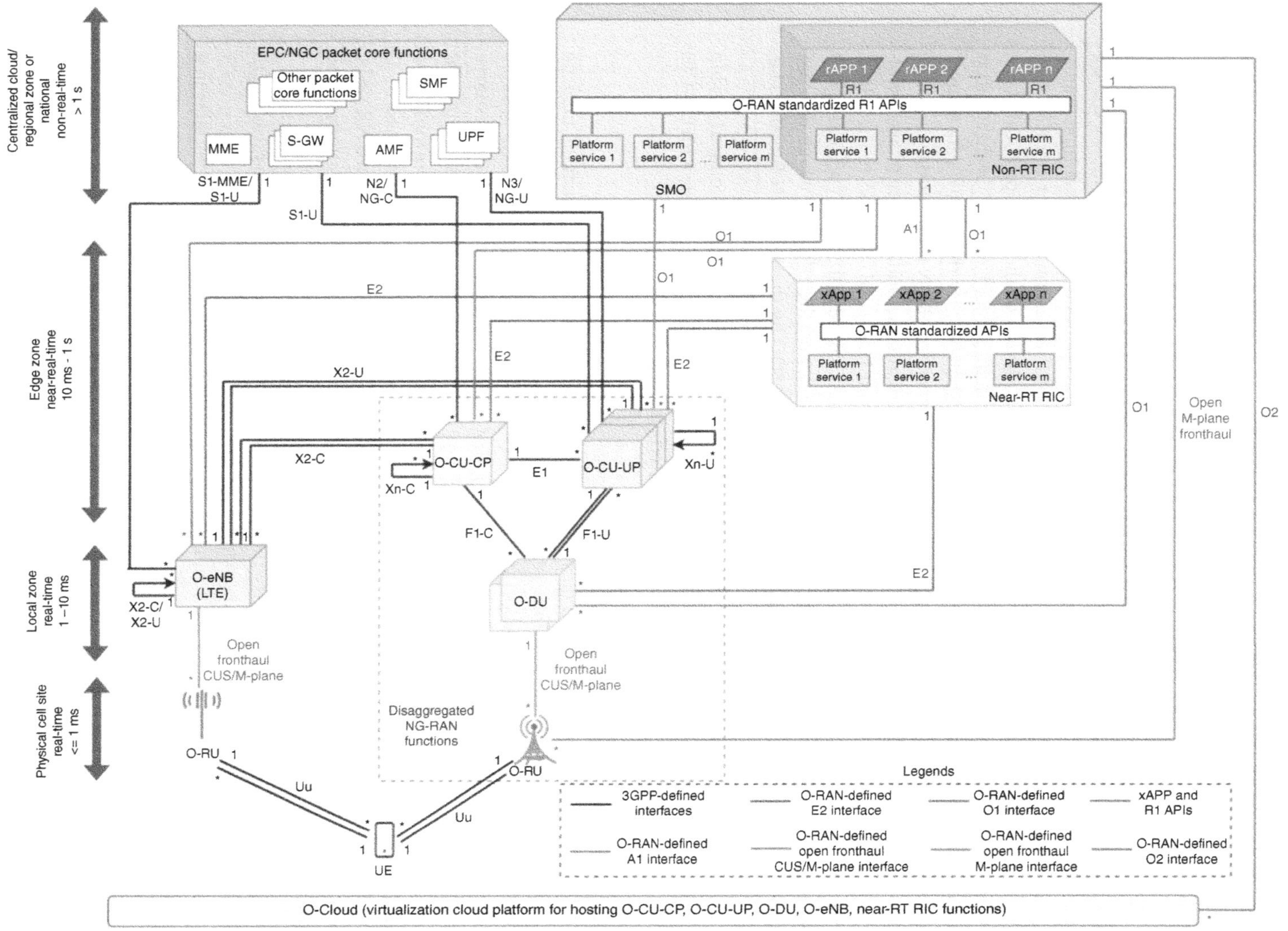

Figure 3.1 An informal view of the O-RAN functions.

In a disaggregated RAN architecture, the baseband unit of a 5G new radio (NR) base station, also called the next generation Node B (gNB), is logically split into a single centralized Unit Control Plane (CU-CP) function, one or more centralized unit user plane (CU-UP) functions, one or more distribution unit (DU) functions. These functions are deployed as logical nodes in the RAN.

3.1.1.1 Centralized Unit – Control Plane and User Plane Functions (CU-CP and CU-UP)

Typically deployed in a cloud platform at the edge of the RAN, the CU functions (i.e. the CU-CP and the CU-UP) manage RAN procedures for a UE that are operated in near-real-time granularities, ranging from 10 ms to 1 s. In particular, the CU-CP is responsible for the RRC and packet data convergence protocol-control plane (PDCP-C) layers of the RAN protocol stack, dealing with connectivity, mobility, signaling radio bearer management, and UE/bearer context establishment RAN procedures for a UE. The CU-UP is responsible for the service data adaptation protocol (SDAP) and PDCP-user plane (PDCP-U) layers of the RAN protocol stack, dealing with the establishment of Data Radio Bearer (DRB) for a UE and the multiplexing of quality-of-service (QoS) flows for the UE-subscribed IP Protocol Data Unit (PDU) session to the DRBs. Each QoS flow of a PDU session has a corresponding 5G QoS Class identification (5QI), assigned by the core network session management functions (SMFs), and the DRBs, to which the QoS flows are multiplexed, are also assigned a QoS profile (i.e. 5QI) by the NG-RAN function, i.e. the CU-CP.

The CU-CP node communicates with the CU-UP nodes using the third-generation partnership project (3GPP)-defined E1 interface. The CU-CP and CU-UPs of a gNB communicate with CU-CPs and CU-UPs of other gNBs using Xn-C and Xn-U interfaces, respectively.

3.1.1.2 Distributed Unit Function (DU)

Deployed in a baseband cloud pool closer to the physical cell site, the DU functions manage RAN procedures for a UE at real-time granularities, ranging from 1 to 10 ms. In particular, the DU is responsible for the (i) Radio Link Control (RLC) layer – dealing with DRB-specific buffer management, RLC segmentation, etc.; (ii) MAC layer – responsible for radio resource allocation, rate adaptation, etc.; and (iii) PHY-U layer – for managing cells and component carriers (CCs), channel estimation, precoding weight assignment, digital beamforming, etc. The DU node communicates with the CU-CP and CU-UP nodes using the 3GPP-defined F1-C and F1-U interfaces, respectively. A DU is responsible for managing one or more NR cells, each corresponding to a CC with a given central band frequency, and their frequency-time resources, namely:

(i) **physical resource blocks** (PRBs) – the minimum unit of allocation of spectrum resources to UEs in the frequency domain – it comprises of 12 subcarriers;
(ii) **bandwidth parts** (BWPs) – the frequency regions within a CC, each of which pertains to a given numerology configuration [in terms of subcarrier spacing/slot duration], that is different from other regions having a different numerology; and
(iii) **transmission time intervals** (TTIs) – the unit of allocation of spectrum resources to UEs in the time domain.

3.1.1.3 Radio Unit Function (RU)

Deployed at the actual physical cell site as a physical network function (PNF), the radio unit (RU) of the NR gNB is responsible for lower PHY (PHY-L) layer procedures, dealing with radio frequency transmission, mapping of logical antenna ports to antenna elements, analog beamforming, etc., at ≤1 ms real-time granularities. A DU node from the baseband communicates with one or more RUs over a fronthaul interface. Each RU transmits one or more NR cells managed by the DU. The UE connects to the RAN via the RU over the Uu interface.

3.1.1.4 Evolved Node B (eNB)

As far as long-term evolution (LTE) is concerned, the baseband unit of a 4G LTE base station, also called Evolved Node B (eNB), is decoupled from the RU. The layers RRC, PDCP, RLC, MAC, and PHY are all managed in a single baseband unit of the eNB, deployed at the baseband pool cloud close to the physical cell site. As in the case of NR, the RU is deployed in the physical cell site and is responsible for lower PHY layer, communicating with the UE over the Uu interface. Similarly, the eNB manages one or more E-UTRA cells and their CCs that are transmitted by the RUs. The eNB communicates with other eNBs using the X2-C and X2-U interfaces for relevant control plane and user plane procedures, respectively. In the case of E-UTRA-NR dual connectivity (EN-DC), also called LTE-NR DC, the eNB communicates with the CU-CP and CU-UP nodes of NR gNBs using X2-C and X2-U, respectively.

O-RAN architecture focuses on openness of the RAN based on the following key principles:

- **Multi-vendor interoperability**: To facilitate interoperability across vendors between different logical nodes and different RAN functions, the interfaces, such as X2, Xn, F1, E1, and fronthaul, are implemented as open interfaces, as per O-RAN specifications.
- **RRM split for RAN optimization**: To build RAN functionalities to realize use cases with service guarantees and slice assurance, the RRM functionalities are split from the RAN and migrated to the RIC functions over new O-RAN-defined open interfaces, such as E2, A1, Y1, and O1, based on software-defined networking principles.
- **Virtualization and cloudification**: The local baseband and edge cloud deployments mean that the RAN functions are virtualized, i.e. the software components of the RAN functions implemented as NFs are migrated from vendor-proprietary hardware to COTS cloud platforms running on white-box hardware, where they are implemented as virtualized network functions (VNFs) or cloud-native network functions (CNFs).
- **Third-party intelligence and app ecosystem**: The RAN functions of CU-CP, CU-UP, DU, eNB, RU are open to RRM by new functions defined in the O-RAN architecture, namely, the RIC functions and the service management and orchestration (SMO) function. To foster a competitive, innovative, and open ecosystem for RAN optimization, dedicated third-party xApps/rApps with RAN optimization solutions leveraging sophisticated AI/ML tools toward controlling RRM decisions for individual RAN functionalities and O-RAN use cases, such as traffic steering (TS), QoS, slicing, resource optimization, and MIMO, can be onboarded onto the RIC platforms. The apps interface with the RIC platform functions using O-RAN-defined open APIs, R1 interface, etc.
- **Cloud orchestration and deployment**: To manage infrastructure resources and the deployment life cycle for vRAN NFs and apps in the cloud using multi-vendor orchestration functions from the SMO over the new O-RAN-defined O2 interface. This helps optimally orchestrate the computational and storage resources for the O-RAN functions in the cloud resource pools.

Disaggregated RAN functions that are compliant with the above O-RAN architecture principles are prefixed with an "O-," such as O-CU-CP, O-CU-UP, O-DU, O-RU, and O-eNB. The formal O-RAN architecture with all integral RAN functions, RIC/SMO functions, and interfaces are shown holistically in Figure 3.2.

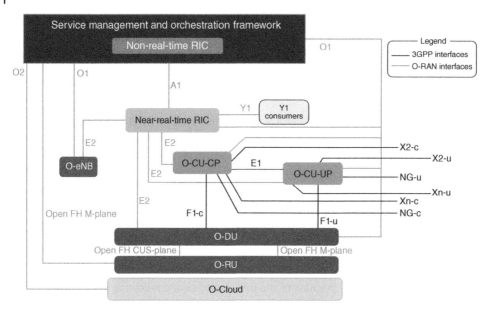

Figure 3.2 O-RAN architecture. *Source:* O-RAN.

3.1.2   RAN Intelligent Controller (RIC) and Service Management and Orchestration (SMO) Functions

As discussed above, intelligence is crucial in O-RAN. The RIC leverages intelligence toward making optimal RRM decisions for RAN functionalities at UE-level, UE group-level, DRB/QoS flow-level, cell-level, slice-level, node-level, etc. There are two RIC functions defined in the O-RAN architecture. One is the near-real-time RAN Intelligent Controller (near-RT RIC), deployed in a cloud infrastructure (usually at the edge of the RAN), which is responsible for fine-grained RRM decisions of control-plane and user-plane procedures (mostly UE-specific) pertaining to functionalities across the RAN protocol stack at near-real-time granularities with low-latency closed-loop control loops ranging from 10 ms to 1 s over the O-RAN-standardized E2 interface.

The other function is the non-real-time RAN Intelligent Controller (non-RT RIC) deployed within the Service Management and Orchestration (SMO) framework function in a centralized cloud. The SMO is responsible (i) for fault, configuration, accounting, performance, and security (FCAPS) operations of the management-plane functions of the RAN and (ii) for the orchestration of cloud infrastructure resources to manage the RAN functions. The non-RT RIC is responsible for configuring high-level declarative policies, setting Key Performance Indicator (KPI) objective targets and generating enrichment information toward closed-loop RRM of individual UEs, UE groups, cells, slices, etc. at non-RT granularities ranging from 1 s and above. The non-RT RIC is also responsible for providing AI/ML services to the near-RT RIC, such as ML model training services, version control, ML model catalog maintenance, ML model repository maintenance, etc., which can be harnessed by the near-RT RIC toward making inference decisions.

The RIC functions exercise closed-loop control of the RAN functions using O-RAN-standardized open interfaces, as detailed below. Extensible applications from third parties, known as xApps and rApps, are onboarded in the near-RT RIC and non-RT RIC functions, respectively. Each xApp/rApp uses custom logic, integrating AI/ML algorithms, to make RRM decisions for a dedicated RAN functionality (such as traffic steering, QoS provisioning, slicing, MIMO, and resource allocation) and interfaces with the near-RT RIC/non-RT RIC platform functions toward exercising the RRM decisions on the RAN functions via closed-loop control actions over O-RAN-defined open interfaces.

3.1.3 Interfaces

X2-C interface: The interface, standardized in 3GPP and adopted in O-RAN, between any two O-eNBs and between an O-eNB and an NG-RAN O-CU-CP, handling relevant control plane signaling procedures. This interface is further discussed in Section 3.5.

X2-U interface: The interface, standardized in 3GPP and adopted in O-RAN, between any two O-eNB and between an O-eNB and an NG-RAN O-CU-UP, handling relevant user plane procedures. This interface is further discussed in Section 3.5.

Xn-C interface: The interface, standardized in 3GPP and adopted in O-RAN, between any two NG-RAN O-CU-CPs, handling relevant control plane signaling procedures. This interface is further discussed in Section 3.5.

Xn-U interface: The interface, standardized in 3GPP and adopted in O-RAN, between any two NG-RAN O-CU-UPs, handling relevant user plane signaling procedures. This interface is further discussed in Section 3.5.

F1-C interface: The interface, standardized in 3GPP and adopted in O-RAN, between an NG-RAN O-CU-CP and an NG-RAN O-DU, handling relevant control plane signaling procedures. This interface is further discussed in Section 3.5.

F1-U interface: The interface, standardized in 3GPP and adopted in O-RAN, between an NG-RAN O-CU-UP and an NG-RAN O-DU, handling relevant user-plane signaling procedures. This interface is further discussed in Section 3.5.

E1 interface: The interface, standardized in 3GPP and adopted in O-RAN, between an NG-RAN O-CU-CP and an NG-RAN O-CU-UP, handling relevant control-plane signaling procedures. This interface is further discussed in Section 3.5.

Open Fronthaul CUS plane interface: The interface, standardized in O-RAN, between an NG-RAN O-DU and an O-RU, for the lower-layer (i.e. the PHY layer) functional split for control plane, user plane, and synchronization operations. This interface is further discussed in Section 3.5.

Open Fronthaul M-plane interface: The interface, standardized in O-RAN, between an NG-RAN O-DU and an O-RU for management plane operations. This interface is further discussed in Section 3.5. Note that the interface between the SMO and the O-RU is also referred to as the open fronthaul M-plane (further discussed next).

Uu interface: The over-the-air interface between an O-RU and a UE, handling relevant signal transmission procedures.

E2 interface and xApp APIs: Connecting the near-RT RIC with the O-CU-CP, O-CU-UP, O-O-DU, and O-O-eNB logical nodes, the E2 interface is bidirectional over which the RRM for the RAN functions is split. The logical nodes are collectively referred to as the E2 nodes. Along with the O-CU nodes, the near-RT RIC is deployed in the edge cloud at the edge of the RAN and is responsible for RRM decisions of RAN procedures operating at near-RT granularities over low-latency control loops. The RRM split basically means that the call processing and signaling procedures are implemented in the E2 nodes, but the RRM decisions for these procedures are controlled by the RIC function. As an example, the HO procedure for a UE is processed by the E2 node, but the decision of which should be the target cell for HO is controlled by the RIC.

The procedures and messages exchanged over the E2 interface are standardized by E2 Application Protocol (E2AP). Using E2AP, the E2 nodes can send reports (containing crucial RAN data or UE context information) to the near-RT RIC and/or request for RRM control from near-RT RIC over E2. In addition, the near-RT RIC can send low-latency control actions (containing RRM decisions), imperative policies, and subscriptions to the E2 node from the RIC via E2AP.

The xApps in the near-RT RIC encode/decode the payload content of the E2AP messages containing RRM-specific information, defined by E2 Service Models (E2SM), where the service models define the semantics of RRM operations over E2. The near-RT RIC platform functions route E2AP messages from/to E2 nodes via O-RAN-defined APIs to/from the xApps. The xApps interface with the near-RT RIC platform functions using standardized xApp APIs.

O1 and A1 interfaces and R1 APIs: While the near-RT RIC focuses on fine-grained UE-level RRM for control-plane and user-plane procedures of the E2 nodes across the RAN protocol stack at near-RT granularities (ranging from 10 ms to 1 s) over low-latency control loops, the non-RT RIC is responsible for setting KPI objective targets and high-level declarative policies, generating configuration recommendations and enrichment information, training offline ML models, and managing ML life cycle toward RRM for RAN functionalities at UE-level, UE group-level, cell-level, slice-level, etc. at non-RT granularities (greater than 1 s). The non-RT RIC is deployed as a service within the SMO function in the centralized cloud and does not directly interface with the E2 nodes and O-RU but via the near-RT RIC and the SMO. The SMO is responsible for FCAPS RRM of management-plane RAN parameters, mostly cell-level and node-level, in the RAN functions and the near-RT RIC over the O1 interface using non-RT control loops. The non-RT RIC communicates with the near-RT RIC over the A1 interface. A non-RT RIC function controls multiple near-RT RIC functions. Like the xApps in the near-RT RIC, the rApps in the non-RT RIC are responsible for non-RT RRM decisions for RAN functionalities. The rApps use APIs defined for the R1 interface to communicate with the non-RT RIC and SMO platform services.

O2 interface: The SMO is additionally responsible for orchestration and management of infrastructure resources in the cloud platforms hosting the RAN and the RIC VNFs over the O2 interface. The cloud platforms are referred to as the "O-Cloud" function in O-RAN architecture. The interfaces are shown in Figures 3.1 and 3.2.

Y1 interface: Y1 is a new interface being defined in O-RAN for exposure of analytics information from Near-RT RIC to authorized consumers. It allows the consumers to subscribe to request the RAN analytics information service(s) provided by Near-RT RIC. The interface is shown in Figure 3.2.

3.2 Near-RT RIC Architecture

Here, we discuss the standard functional architecture of the O-RAN near-RT RIC function and the supported features. The functional architecture principles of the near-RT RIC are shown as follows:

3.2.1 Standard Functional Architecture Principles

- The O-RAN near-RT RIC is a logical function in the O-RAN architecture that enables near-real-time control and optimization of RAN elements, resources, and functions of the E2 nodes via fine-grained data collection and actions over the E2 interface with low-latency control loops in the order of 10 ms–1 s.
- The near-RT RIC and the E2 node functions are fully separated from the transport functions. The addressing scheme used in the near-RT RIC and the E2 nodes shall not be tied to the addressing schemes of transport functions.

- The O-CU-CP, the O-CU-UP, and the O-DU nodes of NG-RAN and the O-eNB are collectively referred to as E2 nodes. A near-RT RIC can be connected to multiple E2 nodes. The E2 interface termination function in the near-RT RIC terminates the E2 interface from an E2 node. One physical network element can implement multiple logical nodes, and the near-RT RIC interfaces are based on the logical model of the entity controlled using the interfaces.
- The E2 nodes support all radio protocol layers (RRC, SDAP, PDCP, RLC, MAC, PHY) and inter- faces (NG, S1, X2, Xn, F1, E1, etc.) defined within 3GPP RANs that include eNB for E-UTRAN and gNB/ng-eNB for NG-RAN.
- The near-RT RIC hosts one or more xApps (third-party apps for RRM) that make use of the E2 interface to collect near-real-time information on a per-UE basis or on a per-cell basis via the near-RT RIC platform functions and provides value-added services by controlling the RAN func- tionalities managed by the procedures of the E2 node on a per-UE basis.
- The set of procedures associated with the E2 interface constitute the E2 Application Protocol (E2AP), which enables a direct association between the xApp and the RAN functionality over E2. The transport network layer for the E2AP interface is built on IP transport, and for the reliable transport of signaling messages, stream control transmission protocol (SCTP) is added on top of IP.
- When configurations with multiple SCTP associations are supported, the near-RT RIC may request to dynamically add/remove SCTP associations between the E2 node and the near-RT RIC pair. The near-RT RIC and the E2 node support multiple transport network layer (TNL) associations over the E2 interface.
- Individual xApps in near-RT RIC may address specific RAN functions in a specific E2 Node. The E2 termination function routes xApp-related messages to the target xApp. The near-RT RIC and the hosted xApps shall use a set of services exposed by the E2 node, described by a series of RAN function-specific "E2 Service Models (E2SM)." These services provide the near-RT RIC with access to messages and measurements exposed from the E2 node (such as cell configuration information, supported slices, PLMN identity, network measurements, and UE context informa- tion) and enable control of the E2 node from the RIC. The E2SM describes the functions in the E2 node that may be controlled by the near-RT RIC and the related procedures, thus defining a function-specific RRM split between the E2 node and the near-RT RIC. The RRM functional allocation between the near-RT RIC and the E2 node is subject to the capability of the E2 node exposed over the E2 interface by means of the E2 service model to support the O-RAN use cases. In addition to control, the near-RT RIC can monitor, suspend, stop, override, or regulate (via policies) the functions of the E2 node.
- The xApps in the near-RT RIC interface with the platform functions of the near-RT RIC using APIs, standardized or being standardized, in O-RAN WG3.
- The near-RT RIC is connected to the non-RT RIC, which is deployed in the SMO framework, through the A1 interface. The non-RT RIC provides the UE-level, cell-level, slice-level, group- level declarative policies, enrichment information, and ML model deployment to the near-RT RIC for RAN optimization. A near-RT RIC is connected to only one non-RT RIC.
- The near-RT RIC is connected to the SMO via the O1 interface. This enables forwarding network management messages from the SMO layer to the near-RT RIC management function.
- The near-RT RIC is deployed at the edge of the RAN, and the non-RT RIC is deployed in the SMO in the centralized cloud. The SMO/non-RT RIC has a large computational and storage capacity to build offline ML models and typically acts as an ML training host. The SMO/non-RT RIC may use a graphical processing unit (GPU). The near-RT RIC, on the other hand, uses a central processing unit (CPU) and is typically deployed colocated with the CU or at the edge of the RAN controlling units of CU-CPs. The near-RT RIC has relatively less computational and storage capacity. It can be used only to build online learning or update of ML models and to act as an inference host.

3.2.2 E2 Interface Design Principles

- The E2 interface is open and supports the exchange of control signaling information between the endpoints. From a logical standpoint, the E2 is a point-to-point interface between the endpoints.
- The E2 interface is based on SCTP transport protocol with ASN.1 encoding format for the messages sent over the interface.
- The E2 interface should also support the ability to provide UE ID information, in order to distinctly identify individual UEs, toward the near-RT RIC based on a pre-configured trigger that could be either periodic or event-driven.
- The E2 node consists of one or more RAN functions that are controlled by the near-RT RIC, which support near-RT RIC services, and other RAN functions that do not support near-RT RIC services. The E2 interface specifications shall facilitate the connectivity between the near-RT RIC and the E2 node supplied by different vendors.
- The near-RT RIC may use the following services provided by the E2 node via its RAN functions that support RIC services
  - **REPORT services**: The E2 node exposes data, such as cell configuration, supported slices, PLMNs, network measurements, and UE context information, to the near-RT RIC via an indication message.
  - **INSERT services**: The E2 node requests for control of RAN functionalities, associated procedures, and configurable parameters to the near-RT RIC via an indication message and suspends the ongoing call processing until it receives a control action from the RIC, following which the E2 node resumes the suspended call processing. While initiating the request via Insert indication, the E2 node may also additionally include cell configuration information, UE context information, network measurements, etc.
  - **CONTROL services**: Here, the E2 node enables the near-RT RIC to control selected functions of the E2 node and the associated RAN parameters. The near-RT RIC can asynchronously send a control action to the E2 node or send it as a response to a previous Insert indication. In the latter case, the E2 node shall resume call processing (that it suspended, following an insert indication) upon receiving the control action and triggers the appropriate procedures.
  - **POLICY services**: The E2 node enables the near-RT RIC to offer a prescriptive guidance to the E2 node on controlling the RAN functions and the values of the associated RAN parameters on the E2 node. The near-RT RIC does not send a control action that directly controls the call processing procedures in the E2 node but offers a guidance to the control logic of the E2 node on how to process the ongoing call based on changes in the network conditions, traffic conditions, UE context, UE and network performance, etc. Such policies are generated and sent asynchronously from the near-RT RIC.
- The RIC services are realized using the following E2AP procedures (O-RAN-WG3.E2GAP):
  - **E2AP RIC subscription procedure**: The near-RT RIC uses this procedure to subscribe to the RIC services provided by the E2 node. The RIC defines the trigger conditions for the subscription of these services, which include both event-driven triggers as well as periodic triggers. The event-driven triggers are defined by a set of condition tests involving associated RAN parameters, their values, and the matching conditions. The periodic triggers are based on predefined timer events. The subscription procedure is also used to specify the subsequent actions that must be executed by the E2 node upon the satisfaction of the trigger conditions. These actions can include:

- **Insert Indication**: where the E2 node sends an indication message to the near-RT RIC containing a request to the RIC to control one or more RAN functionalities, along with one or more associated RAN parameters pertaining to a UE. Note that, as discussed above, an Insert indication requires the E2 node to suspend the ongoing call processing.

  For example when the reference signal received power (RSRP) of the UE with respect to the current serving cell is less than a predefined threshold, the E2 node must request the near-RT RIC if the UE must be handed over to a target cell, and if so, what should be the target primary cell. Up until it hears from the RIC or until a timeout event, the E2 node suspends the ongoing call processing for the UE. Here, the "trigger" is defined by the event involving the UE's RSRP parameter, its value, and the associated matching condition; the "Indication" contains the request coming from the E2 node to the near-RT RIC for handover control of the UE; and the "control parameters" requested by the E2 node is the "target primary cell" for the UE if the RIC decides to do a handover the UE.

- **Report Indication**: where the E2 node sends an indication message to the near-RT RIC containing measurement reports, cell configuration reports, UE context reports, state information, etc. Again, as discussed above, a Report indication does not require suspension of the ongoing call processing at the E2 node, since it is not expecting a response from the RIC to proceed but just sends reports reflecting updates in the UE state, RAN state, etc.

  For example, when there is a modification in the secondary cells configured for a UE following a secondary cell reselection, the E2 node sends the modified UE context from the E2 node to the near-RT RIC. Here, the "trigger" is defined by the event involving secondary cell reselection/modification for the UE, and the "Indication" contains the updated UE context with new secondary cells, which constitute the associated RAN parameters reported by the E2 node. The other example is periodic reporting of a UE throughput by the E2 node to the near-RT RIC. Here, periodic reporting is specified by a timer event defined in the order of near-RT granularities (i.e. tens of milliseconds to one second), say, 50 ms. The "Indication" contains the UE performance measurement reports, and the associated RAN parameters include the UE's DRB throughput.

- **Policy action**: where the E2 node must execute a control action based on the trigger condition and the recommended guidance defined by the near-RT RIC. The E2 node does not raise a request to the near-RT RIC when the trigger condition is satisfied unlike Insert indication but inherently invokes a control action based on the recommended guidance from the near-RT RIC, as defined in the RIC subscription procedure that is performed in the E2 node.

  For example, the policy from the near-RT RIC to the E2 node can be defined such that during the occurrence of an A3 event for a UE (characterized by a neighbor cell yielding more RSRP to the UE than its current serving cell), the E2 node must handover the UE along with its DRBs where the list of DRBs to be admitted by the target primary cell for the UE should be sorted in decreasing order of their allocation retention priority (ARP) values. Here, the "trigger" is defined by the A3 event, the "policy action" is defined by the priority of the DRBs for the UE to be handed over to the target primary cell based on the decreasing order of their ARP values, and the "RAN parameters" include the list of DRBs and their 5QI profiles.

o **RIC indication procedure**: The E2 node uses this procedure to carry the outcome and messages for the RIC indication services of INSERT and REPORT.

o **RIC control procedure**: The near-RT RIC uses this procedure to initiate or modify the RIC CONTROL service.

- Each RAN function supporting RIC services is described in the following terms (O-RAN-WG1, O-RAN-WG3.E2AP):
  - **RAN function definition approach**: Defines the RAN function name and describes the RIC services and the associated RAN parameters that the specific RAN function is currently configured to present over the E2 interface. The E2 node exchanges this information with the near-RT RIC.
  - **RIC event trigger definition approach**: Describes the approach to be used in near-RT RIC subscription messages to set near-RT RIC event trigger definition in the RAN function. The near-RT RIC defines the triggers, which could either be events or periodic timers, in the RIC subscription.
  - **RIC action definition approach**: Describes the approach to be used in subsequent near-RT RIC subscription messages to set the required sequence of near-RT RIC actions in the RAN function. The near-RT RIC also includes the set of associated RAN parameters that must constitute each action.
  - **RIC indication header and RIC indication message approach**: Describes the approach to be used by RAN when composing indication messages for near-RT RIC **REPORT** and **INSERT** services. The E2 node includes the set of associated RAN parameters for each indication message.
  - **RIC control header and RIC control message approach**: Describes the approach to be used by near-RT RIC when composing **CONTROL** messages. The near-RT RIC includes the set of associated RAN parameters for each control message.
  - **RAN function policies**: Describes the set of policies that the RAN function is configured to support and the corresponding parameters that may be used to configure the policy using near-RT RIC policy services. The near-RT RIC includes the set of associated RAN parameters that are part of the policy definition in the RIC subscription.

3.2.3 xApp API Design Architecture

xApps are extensible applications responsible for controlling the RRM of individual RAN functionalities at finer UE-level granularities using low-latency control loops in E2 nodes over the E2 interface based on crucial RAN information received from the E2 nodes. xApps can be from third parties and can interoperate with the near-RT RIC platform functions over standardized APIs, and further over E2 using O-RAN-standard E2AP procedures and messages, using E2 service models. The near-RT RIC architecture to support xAPP APIs is shown in Figure 3.3.

The following are the set of standard platform services and APIs used in the near-RT RIC architecture.

Platform services	APIs	Related API procedures	Purpose
Registration and management services	Management APIs	Management API procedures	To register and authorize xApps and for management of xApps. The platform service is also used to authenticate the xApp, assign an ID to the xApp, etc.
API enablement services	Enablement APIs	API enablement procedures	To enable discovery of APIs by the xApps in order to access platform services
xApp subscription management services	E2 subscription APIs	E2 subscription API procedure, E2 subscription delete API procedure	To enable subscription of RAN data and services by xApps over E2

(Continued)

Platform services	APIs	Related API procedures	Purpose
Conflict mitigation services	E2 guidance APIs	E2 guidance API related procedures	To enable conflict mitigation of RIC services by the xApps
xApp repository services	xApp repository APIs	xApp repository API procedures	To register the xApps in the repository to avail A1 services from the non-RT RIC
Database services	SDL APIs	SDL API procedures	To access R-NIB, UE-NIB, and other namespaces for fetching and storage
OAM services	OAM and FCAPS APIs	OAM and FCAPS API procedures	To access OAM and FCAPS services for the xApps and platform such as MOI creation, modification, etc.
AI/ML training services	AI/ML training APIs	AI/ML training API procedures	To access data pipeline and AI/ML model training services, catalog, repository, etc.
E2 termination	E2 indication/control APIs	E2 indication API and E2 control API procedures	To access indication reports from E2 nodes to xApp, and to send control message from xApp to E2 nodes. Also, to interface endpoint for the E2 interface to send/receive all E2AP messages
A1 termination	A1 related APIs	A1 policy and A1 enrichment information procedures	To enable the xApps in the near-RT RIC to receive policy, enrichment information, AI/ML services from the non-RT RIC over A1. And A1 termination in the near-RT RIC then routes the policy, Enrichment information (EI), and other info to the xApps via the A1 termination APIs

There are standardized procedures for interfacing xApps with platform services over APIs, as mentioned in the above table – we provide few relevant details concerning the procedures.

- Management API procedures:
  a. **xApp registration procedure**: This procedure is used to register the xApp to the platform by passing the relevant info through the API call such as xApp name, software version, YANG schema, fault raised, metrics generated, and supported commands. This procedure is also additionally used for authentication, validation, assignment of ID to xApp, etc.
  b. **xApp deregistration procedure**: This procedure is used to deregister the xApp from the near-RT RIC platform.
  c. **xApp configuration procedure**: This procedure is used to configure an xApp based on a configuration management operation received from the SMO over O1 interface.
- API enablement procedures:
  a. **Near-RT RIC API discovery procedure**: This procedure enables the xApps to discover the near-RT RIC APIs offered by the near-RT RIC platform.
  b. **Event subscription procedure**: This procedure enables the xApps to subscribe to monitor events related to the near-RT RIC APIs.
  c. **Event subscription delete procedure**: This procedure enables the xApps to unsubscribe from events monitoring related to the provided near-RT RIC APIs.
  d. **Event notification**: This procedure shows sending the event notifications to the subscribed xApps based on the trigger event captured at the API enablement.

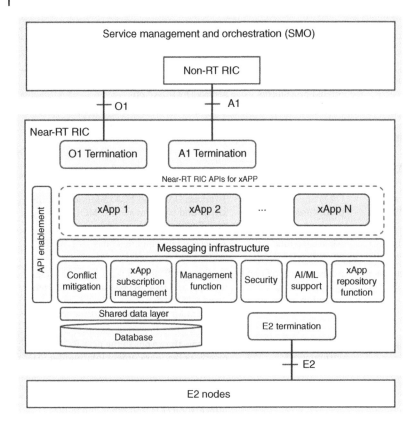

Figure 3.3 Near-RT RIC architecture. *Source:* O-RAN.

- xApp subscription management procedures:
  a. **E2 subscription API procedure**: This procedure is used by the xApps to generate sub-scription to the xApp subscription management service in the near-RT RIC platform that enables the near-RT RIC to make non-duplicate and unique subscriptions to the E2 nodes.
  b. **E2 subscription delete API procedure**: This procedure is used by the xApps to delete an existing subscription from the xApp subscription management service in the near-RT RIC platform.
- E2 termination procedures:
  a. **E2 indication API procedure**: This procedure is used by the xApps to receive RIC INDICATION messages from the E2 node via the E2 Termination.
  b. **E2 control API procedure**: This procedure is used by the xApps to send control action from the xApps to the E2 node via the E2 Termination.
- E2 guidance-related procedures:
  a. **E2 guidance request/response API procedure**: This procedure is used by the xApps to communicate with conflict mitigation services to receive guidance on potential conflicts regarding E2 subscriptions and E2 control.
  b. **E2 guidance modification API procedure**: This procedure is used by the xApps to communicate with the conflict mitigation services to provide a modified guidance to the xApp.

- SDL API procedures:
  a. **SDL client registration procedure**: This procedure is used by the xApps to register with the SDL for permission to access the database.
  b. **SDL client deregistration procedure**: This procedure is used by the xApps to request the SDL to release the registration.
  c. **Fetch data procedure**: This procedure is used by the xApp to request data it is authorized to access from SDL for local processing.
  d. **Subscribe/notify procedure**: This procedure is used by the xApp to subscribe to SDL to be notified of authorized data changes in the database.
  e. **Store data procedure**: This procedure is used by the xApp to insert data into the database.
  f. **Modify data procedure**: This procedure is used by the xApp to update or delete data from the database.
  g. **Subscribe/push procedure**: This procedure is used by the xApp to subscribe to SDL to push authorized and updated data changes in the database to the xApp.

3.3 Non-RT RIC Architecture

Here, we discuss the standard functional architecture of the O-RAN non-RT RIC function and the supported features. Figure 3.4 shows the non-RT RIC architecture. The functional architecture principles of the non-RT RIC are shown as follows.

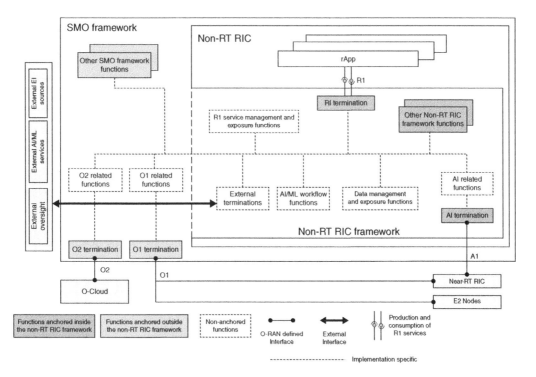

Figure 3.4 Non-RT RIC architecture. *Source:* O-RAN.

3.3.1 Standard Functional Architecture Principles

- The non-RT RIC is a logical function in the O-RAN architecture, present within the Service Management and Orchestration framework (SMO discussed later) that enables non-real-time control and optimization of RAN elements, network elements, resources, and functions.
- The non-RT RIC does not directly interface with the RAN network elements but uses the A1 interface to communicate with the near-RT RIC and communicates with the services of the SMO framework, in which it is deployed (given that the non-RT RIC is a logical function in the SMO). Using the A1 interface, the non-RT RIC sends declarative policies and policy-based guidance of applications at UE-level, UE group-level, cell-level, slice-level, and QoS flow-level, for RRM optimization in the near-RT RIC. The granularity at which non-RT RIC is in the order of >1 s (non-real-time periodicities).
- The SMO is hosted in the centralized cloud. The SMO/non-RT RIC has a large computational and storage capacity to build offline ML models and typically acts as an ML training host. The SMO/non-RT RIC may use a GPU. Hence, the non-RT RIC is responsible for AI/ML workflow including ML model training and updates.
- The non-RT RIC framework has services that are responsible for the overall functioning of the platform, which are responsible for policy services, enrichment information, etc.
- The non-RT RIC hosts one or more rApps (third-party apps for RRM) that make use of the R1 interface to collect data from the network elements or from the near-RT RIC that flow into the SMO via the non-real-time O1 interface, does offline ML model training near-real-time information, and stores ML models in a repository and updates the catalog. These trained models are then deployed in the near-RT RIC via O1, for inference purposes, since near-RT RIC acts as inference host acting on E2 data. These trained models built in the non-RT RIC can also be used for inference on data collected from SMO over R1 to generate AI/ML-trained declarative policies and prediction insights as enrichment information to the near-RT RIC via A1.
- A non-RT RIC can be connected to multiple near-RT RIC functions using the A1 interface. The A1 interface termination function in the non-RT RIC terminates the A1 interface from the near-RT RIC.
- The rApps in the non-RT RIC can communicate over the R1 interface to framework function/services in the non-RT RIC and SMO platforms. The R1 interface termination function in the non-RT RIC terminates the R1 interface at the rApps from the non-RT RIC/SMO framework functions. The R1 interface APIs are standardized or are being standardized in O-RAN WG2.
- The set of procedures associated with the A1 interface constitute the A1 Application Protocol (A1AP), which enables a direct association between the non-RT RIC and the near-RT RIC over A1. The A1 termination function routes rApp-related messages to the target rApp for generating services like A1, EI, etc. Individual rApps in non-RT RIC, associated with a given RRM functionalities (such as traffic steering, QoS, slicing, etc.) may address policies/enrichment information/ML services to individual xApps, associated with the same functionality, in the near-RT RIC.
- The non-RT RIC and the hosted rApps shall use a set of standardized type definitions (and encoding format) for the data types used for generating A1 policies and enrichment information, as standardized in O-RAN WG2. Examples of some of the type definitions for A1 policy include throughput, latency, reliability, packet loss, etc.

3.3.2 A1 Interface Design Principles

- The A1 interface is an open logical interface within the O-RAN architecture between the non-RT RIC functionality in the SMO and the near-RT RIC function, supporting multi-vendor interoperability. It is defined in an extensible way that supports addition of new services and data types without changing the A1 procedures and the application protocol.

- The A1 interface is used to provide the following services:
  a. **A1 policy service**: To provide declarative policies and set KPI targets and objectives for RRM optimization that enable the non-RT RIC function in the SMO to guide the near-RT RIC function in optimizing the RRM for the RAN toward better fulfillment of the RAN intent.
  b. **A1 enrichment information**: To provide information collected or derived at SMO/non-RT RIC from data sources (including non-network data sources) to the near-RT RIC.
  c. **A1 ML model management**: To provide offline ML model training in the non-RT RIC before deployment in the near-RT RIC or the SMO that acts as inference host controlling/configuring the parameters in the RAN functions of the network elements over E2/O1 interfaces.

- The A1 policies are created, modified, and deleted by the non-RT RIC, based on contextual information, and are enforced until deletion. The policies are expressed in standard language and encoding format as standardized in O-RAN WG2. The A1 interface is also bi-directional in the sense that it provides basic feedback from the near-RT RIC that enables non-RT RIC to monitor the use of policies.
  a. The following are the procedures associated with A1 policies:
     i. **Create policy type procedure**: to create an A1 policy at the non-RT RIC.
     ii. **Update policy type procedure**: to update an existing A1 policy at the non-RT RIC.
     iii. **Delete policy type procedure**: to delete an existing A1 policy from the non-RT RIC.
     iv. **Feedback policy procedure**: to provide feedback on the enforcement of the policy from the near-RT RIC to the non-RT RIC over A1.
     v. **Query policy type procedure**: to query the types of A1 policies created by the non-RT RIC.
     vi. **Query policy status procedure**: to query the status of enforcement of an A1 policy from the non-RT RIC to the near-RT RIC.
  b. The A1 policies are enforced at UE-level, UE group-level, cell-level, QoS flow-level, and slice-level, and each of these entities have their respective identifiers to be identified over an A1 policy.
  c. For policy objectives, policy statements have been defined for:
     i. **QoS KPI targets**: setting targets for guaranteed flow bit-rate, maximum flow bit-rate, priority level, packet delay budget, etc.
     ii. **QoE KPI targets**: setting targets for QoE score, initial buffering for video traffic, session rebuffering length for video streaming, video stalling/freezing ratios to the total media length, etc.
     iii. **UE-level targets**: setting targets for UL/DL throughput, UL/DL packet delay, UL/DL packet loss rate at PDCP/RLC layers, and UL/DL reliability.
     iv. **Slice SLA targets**: setting targets for maximum number of UEs, maximum number of PDU sessions, guaranteed DL/UL throughput per UE/slice, Maximum DL/UL throughput per UE/slice, etc.
  d. For policy resources, the policy statements have been defined for traffic steering resources, slice SLA resources, etc.

- The A1 interface enables the near-RT RIC to discover the availability of enrichment information from the non-RT RIC, and for further transmission of A1 enrichment information from non-RT RIC to near-RT RIC, based on requests coming from the near-RT RIC. Enrichment information jobs are created, modified, and deleted by the near-RT RIC based on the RRM objectives for the xApps to manage individual functionalities. The non-RT RIC keeps delivering the enrichment information to the near-RT RIC over A1, until the EI job is deleted.

a. The following are the life cycle aspects of enrichment information:

i. **Registration of enrichment information**: The non-RT RIC sets up the functions that will produce the enrichment information based on the input information made available at the SMO from other data sources, including non-network sources. In the process, an enrichment information type ID (EITypeID) is registered.

ii. **Discovery of enrichment information**: This enables the near-RT RIC to discover enrichment information over the A1 interface, before accessing it.

iii. **Requesting and delivery of enrichment information**: After the near-RT RIC discovers enrichment information over the A1 interface, the near-RT RIC can request to get the EI information by indicating the enrichment information type ID over A1 and by creating an EI job, containing information about the description of EI sought by the near-RT RIC and the delivery mechanism to be used at the non-RT RIC to transmit the information. The non-RT RIC then uses the appropriate mechanism to deliver the A1 enrichment information based on the information on the EI job sent by the near-RT RIC, upon accepting the request for EI from the near-RT RIC.

b. The capabilities and the procedures associated with the A1 enrichment information services are as follows:

i. EI job control:

Create EI job: this procedure is used to request creation of a new enrichment information job by the A1 EI consumer (near-RT RIC) to the A1 EI producer (non-RT RIC) so as to send EI information from the non-RT RIC (A1 EI producer) to the near-RT RIC (A1 EI consumer).

Update EI job: this procedure is used to request update of an existing enrichment information job by the A1 EI consumer (near-RT RIC) to the A1 EI producer (non-RT RIC) to send additional EI information or modify existing EI information being sent from the non-RT RIC (A1 EI producer) to the near-RT RIC (A1 EI consumer).

Delete EI job: this procedure is used to request deletion of an existing enrichment information job by the A1 EI consumer (near-RT RIC) to the A1 EI producer (non-RT RIC) to delete an existing EI job being sent from the non-RT RIC (A1 EI producer) to the near-RT RIC (A1 EI consumer).

Query EI job: this procedure is used to query existing EI jobs by the A1 EI consumer (near-RT RIC) to the A1 EI producer (non-RT RIC) that are currently active for EI information transmission from non-RT RIC (A1 EI producer) to near-RT RIC (A1 EI consumer).

Query EI job status: this procedure is used to query the current status of the EI jobs by the A1 EI consumer (near-RT RIC) to the A1 EI producer (non-RT RIC), to know whether they are currently active or not active, for EI information transmission from non-RT RIC to near-RT RIC.

Notify EI job status: this procedure is used to notify the current status of the EI jobs from the A1 EI producer (non-RT RIC) to the A1 EI consumer (near-RT RIC).

ii. EI discovery:

Query EI identities: this procedure is used to query the EI type identities by the A1 EI consumer (near-RT RIC) to the A1 EI producer (non-RT RIC) to know the list of EI types and their associated identifiers.

Query EI type: this procedure is used to query the EI data type by the A1 EI consumer (near-RT RIC) to the A1 EI producer (non-RT RIC) for getting information about the EI data type from the non-RT RIC (A1-EI producer) to the near-RT RIC (A1-EI consumer).

iii. EI delivery:

Deliver EI result: this procedure is used to deliver the EI information from the A1 EI producer (non-RT RIC) to the A1 EI consumer (near-RT RIC) over A1.

3.3.3 R1 API Design Principles for rApps

rApps are extensible applications responsible for generating declarative policies, KPI targets and setting objectives for RRM of RAN functionalities to the near-RT RIC, and sending recommendations for configuration management of network elements and other FCAPS operations to the SMO. rApps can be from third parties and can interoperate with the non-RT RIC/SMO platform functions over R1 interface APIs, standardized (or currently being standardized) in O-RAN WG2, and further over A1 to the near-RT RIC or O1 to the underlying RAN functions.

The set of R1 services standardized in the non-RT RIC/SMO function are as follows:

SMO/non-RT RIC platform services	Function offering the service	Description and functionality
Bootstrap service	R1 service management and exposure function (SME)	If the bootstrap service is provided by the SMO/non-RT RIC framework, an rApp can use it to discover the endpoints of the service management and exposure services
Service registration and discovery services	R1 service management and exposure function (SME)	To allow service producers to register information about the R1 services they produce and service consumers to discover R1 services they intend to consume
Heartbeat service	R1 service management and exposure function (SME)	A service producer can use the heartbeat service to maintain its R1 service registration status with the service management and exposure services producer if this service is provided
Authentication and authorization	R1 service management and exposure function (SME)	Authentication and authorization are realized as a combination of procedures and services that allows asserting the identity of service producers and service consumers and ensure that only authorized consumers can access the R1 services
rApp registration service	R1 service management and exposure function (SME)	The rApp consumes this service to register with the service management and exposure services producer
Data registration and discovery service	Data management and exposure functions (DME)	Data producers consume this service to register data types. Data consumers consume this service to discover available data types
Data request and subscription service	Data management and exposure functions (DME)	For data consumption, data consumers consume this service to request data instances or subscribe to them, and the data management and exposure functions in the SMO/non-RT RIC framework produce this service. For data collection, the data management and exposure functions in the SMO/non-RT RIC framework consume this service to request data instances or subscribe to them, and data producers produce this service

(Continued)

(Continued)

SMO/non-RT RIC platform services	Function offering the service	Description and functionality
Data delivery services	Data management and exposure functions (DME)	To deliver data to the subscribed/requested data consumers
Network information service	OAM/O1-related functions	The network information service provides to the service consumer information related to the network, in particular the RAN, that has been aggregated from multiple information sources that the SMO has access to, e.g. configuration, topology, network element state, geolocation, inventory, etc.
Performance management service	OAM/O1-related functions	The performance management service allows the service consumer to access performance information that was collected from the network elements by the service producer
Trace service	OAM/O1-related functions	The trace service allows the service consumer to access call trace messages collected from the network elements by the service producer
Configuration management service	OAM/O1-related functions	The CM service allows the service consumer to access configuration information pertaining to the managed entities, as obtained by the CM service producer. The CM service further allows the service consumer to request configuration changes related to the managed entities
AI/ML workflow services	AI/ML workflow functions	This service allows the service producer to offer AI/ML training services and maintenance of AI/ML model repository and the consumer to subscribe to models toward deploying them along with the necessary updates for making inference decisions

The set of R1 interface capabilities that enables processing of the functionalities associated with the R1 services, standardized in O-RAN WG2, is as follows:

SMO/non-RT RIC features	Provisioning service producer	Description and functionality
Bootstrap discovery	R1 SME function	To provision the bootstrap service to the rApp via a fixed well-known address or fully qualified domain name (FQDN).
Registration of services	R1 SME function	A service producer declares its produced services to the service management and exposure services producer
Discovery of services	R1 SME function	For service discovery, a set of procedures enables a service consumer to retrieve information on available services based on selection criteria
Registration status maintenance via heartbeat	R1 SME function	The service producer sends a heartbeat message periodically, where the period may be specified in the response of the register service procedure. On receiving the heartbeat message, the service management and exposure services producer validate it and send a response to the service producer that may include information modifying the heartbeat message period
Authorization of services	R1 SME function	Authorization procedures are used to grant service consumers access to registered R1 services and to allow service producers to produce R1 services

(Continued)

SMO/non-RT RIC features	Provisioning service producer	Description and functionality
rApp registration management	R1 SME function	The rApp consumes this service to register with the service management and exposure services producer and may provide the following information such as rApp name, vendor, software version, certificates, role of the rApp (service producer and/or service consumer), and security credentials. On successful registration, the service management and exposure services producer respond with an rApp identifier (rAppID)
Registration of data types	R1 DME function	To register data types by the data producer with the DME and maintain their registration with the DME
Discovery of data types	DME function	Discovery of a data type does not imply availability of data instances of that data type. It is just an indication that data consumers can request or subscribe data for that data type. It can happen that the actual data production is only triggered by the request or subscription actions of the data consumers
Requesting of data	DME function	This allows a service consumer to specify a data instance (for a registered data type) to be delivered to it and to choose the delivery service to use. They also allow the service producer to notify the service consumer about the availability of data, if such information is applicable for the used delivery service
Retrieval of data	DME function	The pull data service allows the service consumer to obtain data via a pull mechanism from the service producer. Retrieval of data can occur once or multiple times (polling). For data consumption, data consumers consume this service and the data management and exposure functions in the SMO/non-RT RIC framework produce this service. For data collection, the data management and exposure functions in the SMO/non-RT RIC framework consume this service and data producers produce this service
Push data	DME function	The push data service allows the service consumer to push data to the service producer. For data consumption, the data management and exposure functions in the SMO/non-RT RIC framework consume this service and data consumers produce this service. For data collection, data producers consume this service and the data management and exposure functions in the SMO/non-RT RIC framework produce this service
Aggregation of OAM-related data	OAM O1-related function (NIS service)	This shall be used for aggregation of PM data, CM data, topology data, trace data, etc. and provide aggregated data to the consumer
Querying of cell-related information	OAM O1-related function (NIS service)	The network information service supports various information queries that give the rApps access to information aggregated from multiple sources
Querying performance information	OAM O1-related function (PM service)	The PM service shall allow to query performance information using a set of filtering criteria
Configuration management schema retrieval	OAM O1-related function (CM service)	The CM service allows the service consumer to retrieve information pertaining to the configuration schemas of one or more managed entities

(*Continued*)

(Continued)

SMO/non-RT RIC features	Provisioning service producer	Description and functionality
Reading configuration data	OAM O1-related function (CM service)	The CM service allows the service consumer to read the configuration data related to one or more managed entities
Writing configuration changes	OAM O1-related function (CM service)	The CM service allows the service consumer to write configuration changes related to one or more managed entities
Configuration changes subscription notification	OAM O1-related function (CM service)	The CM service allows the service consumer to be notified of CM changes related to one or more managed entities
AI/ML model training and storage	AI/ML workflow function (AI/ML training service)	The AI/ML training service allows the service consumer to provide details on the input features, output variables, choice of algorithm, etc. toward making the platform do training of the ML model
AI/ML model deployment	AI/ML workflow function (AI/ML training service)	The AI/ML training service allows the service consumer to download/deploy the trained model from the platform

The set of R1 interface procedures that enables processing of the functionalities associated with the R1 services, standardized in O-RAN WG2, is as follows:

APIs	Provisioning service producer	Description and functionality
Discover bootstrap service	R1 SME function	This procedure enables determination of the endpoints of the other service management and exposure services
Register service	R1 SME function	A service producer uses this procedure to register an R1 service that it produces with the capabilities that are being exposed. On receiving the request, the service management and exposure services producer determines whether the service producer is authorized to produce the service and whether there are any conflicts. In the response, the service management and exposure services producer informs the service producer on the capabilities that they are allowed to expose
Deregister service	R1 SME function	A service producer can use this procedure to deregister an R1 service that it is registered to provide
Update service registration	R1 SME function	A service producer can use this procedure to update the registration of an R1 service that it provides
Discover services	R1 SME function	A service consumer uses this procedure to discover registered services. The service consumer can provide selection criteria to inquire about the registered services. The response contains a filtered list of available services based on which services the consumer is authorized to access

(Continued)

APIs	Provisioning service producer	Description and functionality
Subscribe service availability	R1 SME function	A service consumer can subscribe to notifications regarding changes in service(s) that are available and thereby receive a notification when a service has been registered, updated, or deregistered by a service producer. The subscription is created for the service consumer based on the services and capabilities that it is authorized to access, i.e. the service consumer only gets notified about changes in registration status of services that it is authorized to access. Upon subscription, the service consumer can pass selection criteria in order to control the set of services about which it wishes to be notified
Unsubscribe service availability	R1 SME function	A service consumer can unsubscribe from notifications regarding changes in the available services
Notify service availability changes	R1 SME function	The service management and exposure services producer can use this procedure to notify a subscribed service consumer about changes in the set of registered services that the service consumer is authorized to access
Heartbeat	R1 SME function	A service producer can use this procedure to maintain its R1 service registrations
Request access	R1 SME function	A service consumer can use the request access procedure to request access to the exposed services that it can discover. A token may be granted for subsequent use of any R1 services that require authorization
Revoke access	R1 SME function	A service consumer with appropriate privileges (administrator) can use the revoke access procedure to revoke the access to exposed R1 services
Register rApp	R1 SME function	An rApp can use this procedure to register with service management and exposure services producer
Register data type	DME function	A data producer uses this procedure to register a data type for which it can produce data instances. On receiving the request, the data registration and discovery service producer determine whether the data producer is authorized to produce the data type and whether there are any conflicts, e.g. if the data type is already registered. In the response, the data registration and discovery service producer informs the data producer if it is allowed to produce the data type
Deregister data type	DME function	A data producer can use this procedure to deregister a data type that it has previously registered for which it is no longer able to produce data instances
Discover data type identifiers	DME function	A data consumer can discover the data types that are available. For each data type, a data type identifier and additional metadata are provided
Query data type information	DME function	A data consumer can retrieve information on a specific data type identified by a data type identifier. Such information enables the data consumer to formulate a request or subscription for data instances of that data type
Subscribe data type changes	DME function	A data consumer can subscribe to notifications regarding changes in the set of available data types and thereby receive notifications when data types have been registered or deregistered. The subscription is created for the data consumer based on the data types that it is authorized to access, i.e. the data consumer can only subscribe to get notified about changes in availability status of data types that it is authorized to access

(*Continued*)

(Continued)

APIs	Provisioning service producer	Description and functionality
Unsubscribe data type changes	DME function	A data consumer can unsubscribe from notifications regarding changes in the set of available data types
Notify data type changes	DME function	The data registration and discovery service producer uses this procedure to notify a subscribed data consumer about changes in the set of available data types
Request data	DME function	The service consumer provides, to the data request and subscription service producer, information on the data it requests based on the data type information
Cancel data request	DME function	The service consumer cancels a data request
Subscribe data	DME function	The service consumer provides, to the data request and subscription service producer, information on the data it wants to subscribe based on the data type information retrieved using the discover data types procedure
Notify data availability	DME function	The service consumer is provided information on the availability and retrieval scheme of the subscribed data instance. As a precondition, the service consumer must have used the subscribe data procedure to create a subscription
Unsubscribe data	DME function	The service consumer cancels a data subscription
Retrieve data	DME function	The service consumer retrieves, from the service producer, data instances about whose existence the service consumer has been informed as part of the request data, notify data availability, or create data offer procedure, using a pull communication mechanism
Push data	DME function	The service consumer provides, to the service producer a data instance using a push communication mechanism as arranged as part of the request data, subscribe data, or create data offer procedure
Cell-related information query	OAM function/ NIS service	This service operation allows to query aggregated information related to RAN cells
CM schema retrieval	OAM function/ CM service	A service consumer can use this procedure to retrieve configuration schemas for the managed entities. The CM service producer will respond with the requested schemas. The schemas provide information about which configuration attributes are supported by the managed entities
CM read configuration	OAM function/ CM service	This procedure enables the service consumer to obtain configuration data (including the configuration attributes) related to one or more managed entities from the CM service producer, subject to optional filtering criteria. The CM service producer responds to the service consumer by providing the requested configuration data or a failure indication along with an appropriate cause
CM write configuration	OAM function/ CM service	This procedure enables the service consumer to request the CM service producer for writing configuration changes related to one or more managed entities. The service producer responds to the service consumer with the status of the write operation for the configuration changes requested by the service consumer or indicate a failure with an appropriate cause

(Continued)

APIs	Provisioning service producer	Description and functionality
CM change subscription	DME function	This procedure enables the service consumer to subscribe to CM changes, if any, on the managed elements
CM change notification	DME function	This procedure enables the service consumer to be notified of CM changes, if any, on the managed elements
PM query information	OAM function/ PM service	This procedure allows to query performance information that has been collected from the network elements. The service consumer specifies a set of filtering criteria to determine the set of information returned

3.4   SMO Architecture

3.4.1   Standard Functional Architecture Principles

Figure 3.5 shows the SMO and non-RT RIC framework to enable rAPPs.

- In a service provider's network, there can be many management domains such as RAN management, core management, transport management, and end-to-end slice management. In the O-RAN architecture, SMO is responsible for RAN domain management. RAN domain management includes managing the vO-CU-CP, vO-CU-UP, vO-DU, vO-eNB, and O-RU network elements; the near-RT RIC; and their associated NFs. The key capabilities of the SMO include:
  (i) FCAPS services include fault management (FM), performance management (PM), configuration management (CM), tracing, heartbeat services, NF discovery, file management, PNF software management, etc. for the E2 node, O-RU, and the near-RT RIC functions. The OAM management with FCAPS services is performed on the E2 nodes and the near-RT RIC from the SMO over the O1 interface and on the O-RU from the SMO over the open M-plane fronthaul interface.
  (ii) Cloudification and orchestration services for the O-Cloud resource pools in terms of compute and storage resources and NF provisioning, life cycle management, etc. for the NFs

Figure 3.5 Exposure of SMO and non-RT RIC framework services. *Source:* O-RAN.

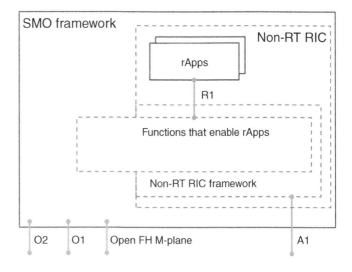

associated with the E2 node, O-RU, and the near-RT RIC. The cloudification and orchestration services are performed on the O-Cloud resource pools (the O-RAN-defined cloud infrastructure) on which the E2 nodes and the near-RT RIC are deployed.

(iii) Offers RAN optimization services from the O-RAN-defined non-RT RIC function to the near-RT RIC.

- The SMO is typically deployed in a centralized cloud (at the regional level or national level) and hosts the non-RT RIC function. There is no O-RAN standard interface, as yet, between the framework functions of the SMO and the non-RT RIC; however, there is an evolving work item in O-RAN, currently, on decomposing the SMO into several functions (SMOFs) that provide one or more SMO services (SMOS), which offer a standardized cohesive set of management, orchestration, and automation capabilities. The SMO functions can interface with each other using standardized interfaces.

- It is to be noted that the SMO framework functions, which provision R1 services, have R1 interface with the rApps in the non-RT RIC. As mentioned above, the SMO has higher storage capabilities for hosting persistent historical information about performance management, configuration management, etc., and may use a GPU for developing offline ML models by the SMO/non-RT RIC framework function.

3.4.2 O1 Interface Design Principles

Some of the key management services supported over the O1 interface include:

- **Provisioning configuration management services**: This allows the SMO (which is the management service MnS consumer) to configure attributes of the O-RAN functions in the managed network elements (the MnS provider, in this case) and to enable reporting of configuration changes from the network elements to the SMO over O1. The attributes of the managed network elements are specified in the information object class (IOC) that define the network resource model for the associated RAN functionalities and network elements. The attributes are managed by the managed object instances (MOIs) of the IOC for the network elements. The associated procedures include:

  a. **Create managed object instance**: This enables the MnS consumer to create an MOI of the IOC associated with the managed element or the managed function in the MnS provider.

  b. **Modify managed object instance attributes**: This enables the MnS consumer to modify the values of the CM attributes of an MOI of the IOC associated with the managed element or the managed function in the MnS provider.

  c. **Delete managed object instance**: This enables the MnS consumer to delete the MOI of the IOC associated with the managed element or the managed function in the MnS provider.

  d. **Read managed object instance attributes**: This enables the MnS consumer to read the values of the CM attributes of an MOI of the IOC associated with the managed element or the managed function, associated with the MnS provider.

  e. **Notify managed object instance attribute value changes**: This enables the MnS provider to notify the MnS consumer of any change in the values of the CM attributes of an MOI of the IOC associated with the managed element or the managed function in the MnS provider.

  f. **Notify managed object instance creation**: This enables the MnS provider to notify the MnS consumer of creation of an MOI of the IOC associated with the managed element or the managed function in the MnS provider.

g. **Notify managed object instance deletion**: This enables the MnS provider to notify the MnS consumer of deletion of an MOI of the IOC associated with the managed element or the managed function in the MnS provider.

h. **Notify managed object instance changes**: This enables the MnS provider to notify the MnS consumer of any changes in the MOI of the IOC associated with the managed element or the managed function in the MnS provider.

i. **Subscription control**: This enables the MnS consumer to create a subscription request to the MnS provider for providing notification about any MOI-specific events for the IOCs associated with the managed elements or the managed functions in the MnS provider.

j. **NETCONF session establishment/termination**: This procedure is used to establish the NETCONF session between the MnS producer and the MnS consumer.

k. **Lock/unlock data store**: This procedure is used by the MnS consumer to lock a target data store on the MnS provider to prevent conflicting writes on the target data store from other functions.

l. **Commit**: This procedure is used by the MnS consumer to commit the configuration changes to the MnS provider of the network element so that the changes are written from the candidate data store to the running data store of the MnS provider.

- **Performance assurance management services**: These services are responsible for the file-based reporting and/or streaming of PMs from the O-RAN functions in the underlying O-RAN network elements (MnS provider) over O1 to the SMO (MnS consumer). Similar to configuration management services, MOIs are created corresponding to the IOCs for the O-RAN functions in the network elements, and these MOIs are responsible for job control of PM jobs and file-based reporting/streaming of PMs from the network elements. The associated services include:

a. PM streaming services:
    i. Establish streaming connection operation to stream data from MnS provider to MnS consumer.
    ii. Terminate streaming connection operation between an MnS provider and MnS consumer.
    iii. Report stream data operation to stream data from MnS provider to MnS consumer.
    iv. Add stream operation to add streams on an existing connection.
    v. Delete stream operation to delete streams from an existing connection.
    vi. Fetch connection info operation to enable the MnS consumer to fetch information pertaining to the streaming connection from the MnS provider.
    vii. Fetch stream info operation to enable the MnS consumer to fetch stream info from the MnS provider.

b. PM data file reporting services:
    i. File ready notification to inform the MnS consumer of the readiness of the file for PM reporting or trace reporting from MnS provider.
    ii. Retrieve PM files to enable the retrieval of PM data stored in files from the MnS provider.

c. Measurement job control services:
    i. **Create PM job**: This procedure is used to create a PM job over O1 with the required set of PMs, KPIs, the supported measurement granularity, the supported reporting periodicity, and the supporting mechanisms (file/stream).
    ii. **Delete PM job**: This procedure is used to delete an existing PM job over O1

d. Threshold crossing notification services:
    i. **Notify threshold crossing**: This procedure is used to notify the MnS consumer of PM threshold crossing by the MnS provider.

- **Fault supervision management services**: These services are responsible for reporting errors and events to the SMO. The SMO performs fault supervision operations on the O-RAN functions in the underlying network elements. The associated procedures include:
  a. Fault notification procedures:
     i. Notify new alarm
     ii. Notify changed alarm general
     iii. Notify cleared alarm
     iv. Notify alarm list rebuilt
     v. Subscribe/unsubscribe to network events
     vi. Get alarm list
     vii. Notify correlation notification changed
     viii. Get alarm count
  b. Fault supervision control procedures:
     i. Acknowledge/unacknowledge alarms
     ii. Clear alarms
     iii. Notify cleared alarms
     iv. Notify ACK state changed
     v. Notify potential faulty alarm list
     vi. Notify changed alarm general
- **Trace management services**: These services are responsible for file-based reporting or streaming of trace records from the O-RAN functions in the network elements to the SMO. Tracing provides the capability to log data on any interface at call level for a given user or mobile type or a given cell, which cannot be deduced from PM data, since PMs provide aggregated values for a given observation period. This granular information is used for going deeper into investigation in the case of anomalous events in the network, compared to PMs that are mandatory for daily operations, future network planning, primary troubleshooting, etc. The different types of tracing services include:
  a. **Cell-level call tracing**: This is used to fetch network traces for call signaling messages, involving a given cell. There are different levels of details associated with call tracing, such as:
     i. **Minimum**: retrieving a decoded subset of information elements (IEs) in the signaling interface
     ii. **Medium**: retrieval of a decoded subset of IEs in the signaling interface along with a selected set of decoded radio measurement IEs
     iii. **Maximum**: retrieval of the entire signaling interface messages with the trace scope in encoded format. The SMO can control which level of tracing detail it seeks from the underlying network element.

     The procedures associated with call tracing include:
     i. Trace data reporting
     ii. Trace session activation/deactivation
     iii. Trace recording session activation/deactivation
  b. **Minimization of drive testing (MDT)**: This is more of a UE-level trace, and there are two types of MDT functionality:
     i. **Immediate MDT**: These are tracing measurements performed by the UE when it is in RRC_CONNECTED state and the measurements are reported to the RAN at the time of availability of reporting condition as well as measurements. Immediate MDT may be file-based or streaming-based.
     ii. **Logged MDT**: These are tracing measurements performed and logged by the UE when it is in RRC_IDLE or RRC_INACTIVE (5G) state, and then, when the UE gets to

RRC_CONNECTED state, the logged measurements are uploaded to the network. Logged MDT are always file-based.

   iii. The list of measurements in MDT include DL signal quantity measurements, power headroom measurement, data volume measurement for DL/UL per DRB per UE, average UE throughput measurement separately for DL/UL and per DRB per UE, packet delay measurement separately for DL/UL and per DRB per UE, etc.

   iv. MDT also allows configuring the reporting trigger, report interval, report amount, event threshold for RSRP/RSRQ, logging interval, logging duration, collection period, etc.

c. **Radio link failure (RLF)**: This capability includes tracing the radio link failures of the UEs with the RAN, along with indication of the appropriate causes.

d. **RRC connection establishment failure**: This capability includes tracing the RRC connection establishment failures of the UEs with the RAN, along with indication of the appropriate causes.

e. **Trace control**: This capability includes the ability of the MnS consumer to create trace jobs for call trace, MDT trace, RLF and RRC connection establishment failure (RCEF) traces, along with necessary job control parameters.

3.4.3 Open M-Plane Fronthaul Design Principles

The key services associated with the open M-plane fronthaul to enable SMO to configure O-RU include:

- Configuration management services
  a. Retrieve state of the O-RU at the SMO, which includes the admin state (locked, shutting down, and unlocked states) and the power state (awake, sleeping states) of the O-RU.
  b. Modify the admin-state and power-state of the O-RU by the SMO.
  c. Retrieve the configuration parameters of the O-RU at the SMO.
  d. Modify the configuration parameters of the O-RU at the SMO.
  e. Delete the configuration parameters of the O-RU, including both the configuration parameters in the candidate or the running data store.
  f. Notification of updates or events on O-RU configuration changes to the consumers.
- Performance management services
  a. Measurement activation and de-activation, which includes activating the measurement data collection from the O-RU at the consumer or de-activating the measurement data collection from the O-RU at the consumer.
  b. Collection and reporting of measurement result, which includes data collection and reporting of measurement results from the O-RU to the consumer.
- Fault management services
  a. Alarm notifications
  b. Manage alarms request
  c. Fault sources
- File management services
  a. File management operation to upload the files containing the performance measurement and other relevant data from the O-RU to the file management services consumer.
  b. File management operation to retrieve a list of files collected at the O-RU by the SMO for performance measurement file reporting, traces, and other relevant data.
  c. File management operation to download configuration files from the SMO to the O-RU.

The FTP or SFTP protocol is used for file management, upload, and download operations between the O-RU and the SMO over the open M-plane fronthaul interface.

3.4.4 O2 Interface Design Principles

- The O2 is an open logical interface within the O-RAN architecture for communication between the SMO and O-Cloud for management of O-Cloud infrastructure and the deployment life cycle management of O-RAN cloudified NFs that run on O-Cloud.
- The interface is defined in an extensible way that enables new information or functions to be added without necessarily changing the protocol or procedures. This interface enables a multi-vendor environment and is independent of specific implementations of SMO and O-Cloud.
- The O-Cloud consists of multiple Deployment Management Services (DMS), which are the logical services provided by the O-Cloud for managing the life cycle of deployments using cloud resources. Each DMS can manage leased resources from multiple resource pools and span multiple locations.
- The Infrastructure Management Services (IMS) are logical services provisioned by the O-Cloud, providing the interface to orchestrate O-Cloud life cycle processes with the NFs it may host along with other operational procedures. There is a single IMS for O-Cloud that manages all resources of DMS and the resources that are not allocated to any DMS in the O-Cloud.
- The functions to be performed over the O2 interface include:
  a. O-Cloud infrastructure resource management
  b. Managing abstracted resources and DMS
  c. NFs and services deployment orchestration

The O-Cloud infrastructure inventory and the logical clouds where the managed functions are deployed are shown in Figures 3.6 and 3.7, respectively.

- O-RAN clouds are described as a distributed cloud composed of O-Cloud pools, where each pool is a collection of O-Cloud nodes, which are computational resource designators.
- The cloud is divided into the following three planes, namely, the management plane, the control plane, and the deployment plane.
- The SMO shall be able to correlate managed element telemetry to infrastructure and deployment telemetry to aggregate problems to a root cause.
- The O-Cloud shall be able to make all configuration data and any external changes to it available to the SMO.
- O-Cloud telemetry shall minimally consist of fault, performance, and configuration data.
- The SMO shall be able to correlate a managed element to its deployment components.

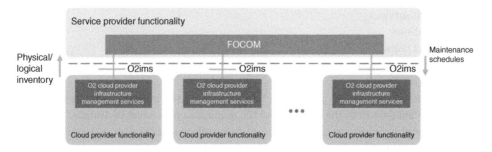

Figure 3.6 O-Cloud infrastructure inventory. *Source:* O-RAN.

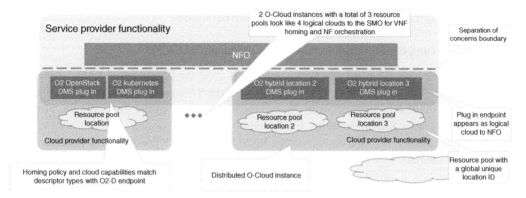

Figure 3.7 Logical clouds. *Source:* O-RAN.

- The O-Cloud shall be able to report telemetry of deployment resources relative to those identified in the deployment descriptor.
- The O-Cloud shall be able to report infrastructure telemetry and identify the deployments using the resource.
- O-Cloud shall provide the collection and reporting of performance information of O-Cloud resources.
- O-Cloud shall support the capability to notify about the availability of performance information.
- O-Cloud shall expose the type of performance information that can be collected for the allocated O-Cloud resources.
- O-Cloud shall expose the type of O-Cloud resource, for which the performance information can be collected.
- O-Cloud shall provide the collection of fault information for O-Cloud resources.
- O-Cloud shall support providing notification of fault information related to O-Cloud resources.
- O-Cloud provisioning shall provide query of O-Cloud capacity.
- O-Cloud provisioning shall provide query of O-Cloud availability.
- O-Cloud shall provide add software images of O-RAN cloudified NF to O-Cloud repository.
- O-Cloud shall provide delete software images of O-RAN cloudified NF from O-Cloud repository.
- O-Cloud shall provide update software images of O-RAN cloudified NF to O-Cloud repository.
- O-Cloud shall provide query software images of O-RAN cloudified NF from O-Cloud repository.
- O-Cloud shall provide software image properties information of O-RAN cloudified NF, such as *softwareImageId*, vendor, and vision.
- O-Cloud life cycle management will provide the (i) deploy, (ii) registration, and (iii) scale capabilities. The objective of deployment is to provide automated provisioning of the O-Cloud infrastructure, while the objective of registration is to register an O-Cloud toward making it available for deployments. Scaling capability is used to scale functional behavior and resources of O-RAN-cloudified NFs to support the required RAN services.
- O-Cloud supports deploying an O-RAN cloudified NF instance.
- O-Cloud supports terminating an O-RAN cloudified NF instance.
- O-Cloud supports horizontal scaling (in and out) of an O-RAN cloudified NF instance.
- O-Cloud supports healing of an O-RAN cloudified NF instance (under study).
- O-Cloud supports querying information about an O-RAN cloudified NF instance.
- O-Cloud supports querying status of life cycle management (LCM) operations.
- O-Cloud supports upgrading of any or all components of an O-RAN cloudified NF instance.

3.5 Other O-RAN Functions and Open Interfaces

3.5.1 O-RAN compliant Centralized Unit Control Plane (O-CU-CP)

As discussed in Section 3.1, this O-RAN function O-CU-CP is the CU-CP of the RAN, responsible for managing the RRC and PDCP-control plane layers of the RAN protocol stack. The RRC layer is responsible for maintaining the connectivity and mobility of users in the RAN, and the PDCP-C layer is responsible for RAN procedures, such as configuring the signaling radio bearers and UE/bearer context establishment for the UEs in the RAN. The O-CU-CP uses the F1-C interface with the O-DU, E1 interface with O-CU-UP, Xn-C interface with other O-CU-CP nodes, X2-C interface with O-eNB, and NG-C interface with the AMF.

3.5.1.1 Control Plane Procedures

The O-CU-CP is responsible for the following control plane procedures, namely:

  (i) Xn-C network interface procedures with other O-CU-CPs for mobility management and multi-RAT dual connectivity procedures such as secondary node addition, modification, deletion, and change.
 (ii) X2-C network interface procedures with E-UTRA O-eNBs for LTE-NR EN-DC dual connectivity procedures.
(iii) F1-C network interface procedures with O-DUs for UE context management (involving primary and secondary cell selection, DRB handover, etc.) and RRC transfer procedures.
 (iv) E1 network interface procedures with O-CU-UPs for bearer context management (involving multiplexing of QoS flows to DRBs, etc.), PDCP SN status transfer, etc.
  (v) NG-C/N2 network interface procedures for initial context management procedures, PDU session establishment, modification, etc., with the NG core access and mobility management function (AMF).

3.5.1.2 Management Plane Procedures

The O-CU-CP can produce performance management, configuration management, and fault management MnS services.

Performance management involves generation of performance metrics related to, say, RRC connection attempts, successful RRC connection attempts, RRC failures, number of handovers, abnormal handovers, etc.

Configuration management involves configuring the parameters associated with the O-CU-CP. Other services include tracing that enables call tracing of the above interfaces, etc.

3.5.2 O-CU-UP

This O-RAN function is the CU-UP of the RAN, responsible for managing the SDAP and PDCP-user plane layers of the RAN protocol stack, which deal with the establishment of DRBs for a UE and the multiplexing of QoS flows for the UE-subscribed IP session to the DRBs. The O-CU-UP uses the NG-U interface to communicate with the user plane function (UPF) and/or the S1-U interface to communicate with the S-GW, over which it receives data packets from the NG core and/or evolved packet core (EPC) core, respectively. Typically deployed in an edge data center O-Cloud platform at the edge of the RAN, the O-CU-CP and O-CU-UP functions manage the RAN procedures for a UE that are operated in near-real-time granularities, ranging from 10 ms to 1 s.

3.5.2.1 Control Plane Procedures

(i) E1 network interface procedures with O-CU-UPs for bearer context management (involving multiplexing of QoS flows to DRBs, etc.), PDCP SN status transfer, etc.

3.5.2.2 User Plane Procedures

 (i) F1-U network interface procedures with O-DUs for transfer of downlink user data, downlink data delivery status, and transfer of assistance information.

(ii) NG-U/N3 network interface procedures with the NG core UPF for transfer of downlink user data, etc.

3.5.2.3 Management Plane Procedures

The O-CU-UP can produce performance management, configuration management, and fault management MnS services.

Performance management involves generation of performance metrics related to, say, PDCP SDU bit-rate, PDCP SDU traffic data volume, PDCP SDU packet delay, etc.

Configuration management involves configuring the parameters associated with the O-CU-UP. Other services include tracing that enables call tracing of the above interfaces, etc.

3.5.3 O-DU

Deployed in a Local Data Center cloud or baseband pool closer to the physical cell site, the O-DU functions manage the RAN procedures for a UE at real-time granularities, ranging from 1 to 10 ms, as discussed in Section 3.1. In particular, the O-DU is responsible for the (i) RLC layer – dealing with QoS bearer-specific buffer management, segmentation, etc.; (ii) medium access control (MAC) layer – responsible for radio resource allocation, rate adaptation, etc.; and (iii) upper PHY (PHY-U) layer – for managing cells and CCs, channel estimation, precoding weight assignment, digital beamforming, etc. The O-DU node communicates with the O-CU-CP and O-CU-UP nodes using the 3GPP-defined F1-C and F1-U interfaces, respectively. A O-DU is responsible for managing one or more NR cells, each corresponding to a CC with a given central band frequency, and their frequency-time resources, namely physical resource blocks (PRBs), BWP, and transmission time interval (TTIs) and allocation of these frequency-time resources to the UEs.

3.5.3.1 Control Plane Procedures

 (i) F1-C network interface procedures with O-CU-CPs for UE context management (involving primary and secondary cell selection, DRB handover, etc.) and RRC transfer procedures.

(ii) The fronthaul interface for control plane operations with the O-RU for downlink control information.

3.5.3.2 User Plane Procedures

 (i) F1-U network interface procedures with O-CU-UPs for transfer of downlink user data, downlink data delivery status, and transfer of assistance information.

(ii) The fronthaul interface for user plane operations with the O-RU for physical data shared channel (PDSCH) data transfer.

3.5.3.3 Management Plane Procedures

The O-DU can produce performance management, configuration management, and fault management MnS services.

Performance management involves generation of performance metrics related to, say, DRB throughput, PRB utilization, number of active UEs, number of transport blocks with different modulation rates, etc.

Configuration management involves configuring the parameters associated with the O-DU. Other services include tracing that enables call tracing of the above interfaces, etc.

3.5.4 O-eNB

As far as LTE is concerned, the baseband unit of a 4G LTE base station, also called O-eNB, is decoupled from the O-RU. The layers RRC, PDCP, RLC, MAC, and PHY-U are all managed in a single baseband unit of the O-eNB, deployed at the local data center (LDC). As in the case of NR, the O-RU is deployed in LDC and is responsible for PHY-L, communicating with the UE over the Uu interface. Similarly, the O-eNB manages one or more E-UTRA cells and their CCs that are transmitted by the O-RUs. The O-eNB communicates with other O-eNBs using the X2-C and X2-U interfaces for relevant control plane and user plane procedures, respectively. In the case of EN-DC, the O-eNB communicates with the O-CU-CP and O-CU-UP nodes of NR O-gNBs using X2-C and X2-U, respectively. The O-eNB also communicates with the mobility management entity (MME) over the S1 interface over which it communicates with the EPC core for maintaining the connectivity and mobility of UEs in the network.

The O-eNB is responsible for the following control plane procedures, namely:

(i) X2-C network interface procedures with NG-RAN O-CU-CPs for LTE-NR EN-DC dual connectivity procedures and with other E-UTRA eNBs for mobility management, dual connectivity operations, etc.
(ii) S1-MME network interface procedures for initial context management procedures, E-RAB establishment, etc., with the EPC packet core's MME.
(iii) S1-U network interface procedure for transfer of IP data from the S-GW to the O-eNB.

3.5.5 O-RU

Deployed at the actual physical cell site, the O-RU of the NR O-gNB is responsible for lower Physical (PHY-L) layer procedures, dealing with radio frequency transmission, mapping of logical antenna ports to antenna elements, analog beamforming, etc. at less than 1 ms granularity, as discussed in Section 3.1. A O-DU node from the baseband communicates with one or more O-RU over a fronthaul interface. Each O-RU transmits one or more NR cells managed by the O-DU. The UE connects to the RAN via the O-RU over the Uu interface.

- The O-RU has control plane (for control plane L1/L2 signaling operations), user plane (such as payload transmission), management plane (such as state management), and synchronization plane interface with the O-DU, called the Open Fronthaul CUS/M-plane interface. The open fronthaul CUS interface also talks about the lower-layer (PHY layer) split in terms of which the O-RAN specifications standardize which PHY level functionalities (called upper-PHY) are assigned to the O-DU and which PHY level functionalities (called lower-PHY) are assigned to the O-RU. The O-RAN specifications of the lower-layer split for the fronthaul interface are further detailed in Chapter 6.
- **The Uu interface**: This interface is responsible for over-the-air transmission between the O-RU and a UE. This is useful for transmitting the RRC messages between the O-RU and the UE and

for data symbol and control signal transmission/reception over the air using the antenna elements.

- The O-RU also has the open fronthaul M-plane interface with the SMO for FCAPS and OAM operations. In O-RAN, there are two modes of operation for the M-plane:
  - **Hierarchical M-plane**: Here, the SMO does *not* directly communicate with the O-RU, but it communicates with the O-DU for the O-RU-specific M-plane operations, and O-DU further performs M-plane operations on the O-RU using the open fronthaul CUS/M-plane interface between the O-DU and O-RU.
  - **Hybrid M-plane**: Here, the SMO directly communicates with the O-RU for M-plane operations over the open fronthaul M-plane interface.
- It is to be noted that important functionalities such as analog beamforming for MU-MIMO and massive MIMO are carried out using the M-plane interface from the SMO to the O-RU.

3.6 Conclusion

In this chapter, we broadly discussed the O-RAN architecture and the functions of the O-RAN architecture. We further discussed the interfaces, standardized in O-RAN alliance, between the O-RAN functions. We provided an in-depth overview of the key procedures associated with the functions of the O-RAN architecture. We also provided details about the capabilities of the O-RAN NFs and the associated services provisioned by the O-RAN functions.

Bibliography

O-RAN-WG1 (2022). *Use-Cases-Detailed-Specification: "O-RAN Working Group 1 Use Cases Detailed Specification"*.

O-RAN-WG1 (2022). *O-RAN-Architecture-Description: "O-RAN Architecture Description"*.

O-RAN-WG3.E2GAP (2022). *O-RAN Working Group 3, Near-Real-time RAN Intelligent Controller, E2 General Aspects and Principles*.

O-RAN-WG3.E2AP (2022). *O-RAN Working Group 3, Near-Real-time RAN Intelligent Controller, E2 Application Protocol (E2AP)*.

O-RAN-WG3-RICARCH (2022). *O-RAN Working Group 3, Near-Real-time RAN Intelligent Controller, Near-RT RIC Architecture*.

O-RAN-WG2-Non-RT-RIC-ARCH-TS (2022). *O-RAN Working Group 2, Non-Real-time RAN Intelligent Controller, Non-RT RIC Architecture*.

O-RAN-WG2-A1AP (2022). *O-RAN Working Group 2, Non-Real-time RAN Intelligent Controller A1 Application Protocol*.

O-RAN-WG2-R1GAP (2022). *O-RAN Working Group 2, Non-Real-time RAN Intelligent Controller R1 General Architecture and Principles*.

O-RAN.WG4.MP (2022). *O-RAN Working Group 4, Management Plane Specification*.

O-RAN.WG4.CUS (2022). *O-RAN Working Group 4, Control User and Synchronization Plane Specification*.

O-RAN.WG5.C.1 (2022). *O-RAN Working Group 5, O-RAN Open F1/W1/E1/X2/Xn Interfaces Working group, NR C-Plane Profile*.

O-RAN.WG5.U.1 (2022). *O-RAN Working Group 5, O-RAN Open F1/W1/E1/X2/Xn Interfaces Working group, NR U-Plane Profile.*

O-RAN.WG6.O2.GAnP (2022). *O-RAN Working Group 6, O-RAN O2 Interface General Architecture and Principles.*

O-RAN.WG10.O1-Interface (2022). *O-RAN Working Group 10, O-RAN Operations and Maintenance Interface Specification.*

O-RAN.WG10.OAM-Architecture (2022). *O-RAN Working Group 10, O-RAN Operations and Maintenance Architecture.*

4

Cloudification and Virtualization

Padma Sudarsan[1] and Sridhar Rajagopal[2]

[1] *VMWare, Palo Alto, CA, USA*
[2] *Mavenir, Richardson, TX, USA*

4.1 Virtualization Trends

Growing "virtualization" trend has been driven by commercial-off-the-shelf (COTS) hardware savings and the potential of delivering applications with the special characteristics of agility, upgrades without service impact, scalability, and resiliency. Operational simplicity and efficiency, time to market, and improving user experience are other key drivers for the transformation.

Service providers want to improve flexibility and deployment velocity, while at the same time reduce the capital and operating costs through the adoption of cloud architectures with flexibility to deploy/lease or reuse hardware. Cloud deployment strategy is generally to "Build Once, Deploy and Manage on Any Cloud," and more and more operators are choosing multi-cloud cloud options and associated platform services. Service providers are transforming their operations, business processes, and infrastructure moving to public, private, and hybrid cloud deployments (Figure 4.1).

While cloud deployments have matured for the wireless core network functions, the radio access network (RAN) functions are still in their infancy. The rest of this chapter is focused on cloud-based deployment for the RAN.

4.2 Openness and Disaggregation with vRAN

Third-generation partnership project (3GPP)/O-RAN has defined an NG-RAN where the existing baseband unit (BBU) is disaggregated into functional components, a radio unit (RU), distributed unit (DU), and central unit (CU) (Figure 4.2).

Conforming to control and user plane separation (CUPS) constructs, the CU can be further decoupled into distinct control plane (CU-CP) and user plane (CU-UP) functions. The CU processes non-real-time protocols and services, and the DU processes PHY and MAC level protocol and real-time services. There are several ways in which the 3GPP layers can be split between the CU, DU, and RU, and generally Layer 2 (L2) non-real-time and Layer 3 (L3) functions are in the CU, Layer 1 (L1) and L2 real-time functions are in the DU, and parts of L1 functions are in the RU.

Open RAN: The Definitive Guide, First Edition. Edited by Ian C. Wong, Aditya Chopra, Sridhar Rajagopal, and Rittwik Jana.
© 2024 The Institute of Electrical and Electronics Engineers, Inc. Published 2024 by John Wiley & Sons, Inc.

Platform economies

- Software running on generic hardware.
- Avoid Vendor Lock-in
- Dimension on network level, not individually per application
- CAPEX benefits through higher utilization levels of assets, aggregation gains and simplified hardware inventories
- Reduced cost through containerization and community standards.

Operational efficiencies

- Grow capacity as needed, pay as you grow
- Automated software upgrades, recover automatically from failures
- OPEX improvements through reduced manual labor
- Improved resiliency and better customer experience

Business agility

- Enabling new services & revenue streams. Fail fast, learn faster.
- Faster release pace
- Programmability and co-creation through open APIs
- Serve new customer segments

Figure 4.1   Virtualization trends.

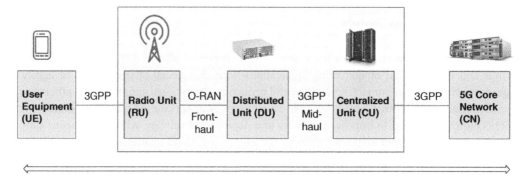

Figure 4.2   Decomposition of the RAN.

The "horizontal" desegregation of the network facilitates the transformation of the next-generation network by enabling flexible and services driven placements of functions.

When we discuss cloud-based deployments, there are three distinct "vertical" disaggregation to consider: *IaaS* (hardware infrastructure) and *CaaS/PaaS* (cloud stack and platform services) that support *Application* (RAN or non-RAN applications).

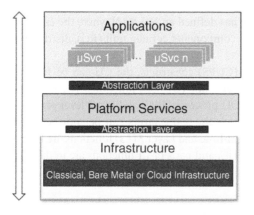

Each layer can come from a different supplier. The first aspect of decoupling has to do with ensuring that a Cloud Stack can work on multiple suppliers' hardware, i.e. it does not require vendor-specific hardware. The second aspect of decoupling has to do with ensuring that a Cloud Platform can support RAN virtualized functions from multiple RAN software suppliers.

Ideally vendors would like to provide the e2e solution (infrastructure, platform services including container services and applications) pre-integrated and custom. However, the reality is, in a cloud-based deployment, that there will be a mix and match of these layers from different vendors. The webscalers (Google Anthos/GKE, Amazon EKS, Microsoft Azure/AKS) and other players like VMware, RedHat, and Wind River dominate this space. Fundamental choice in architecture direction will most likely be driven by service providers, and the reality of the situation is that vRAN applications should be compatible and deployable on several variants of the accelerator in a multi-cloud environment. This is necessitated also by the growing trend to reuse/repurpose/share existing hardware.

The "vertical" and "horizontal" disaggregation of the stack allows a clear separation of the application from the infrastructure and platform services. It provides the necessary transformation needed for the agility, flexibility, and cost savings anticipated with moving to the cloud.

4.3 Cloud Deployment Scenarios

4.3.1 Private, Public, and Hybrid Cloud

The O-Cloud hardware includes compute, networking, and storage components and may also include various acceleration technologies required by the RAN network functions to meet their performance objectives. The O-Cloud software exposes open and well-defined interfaces that enable the orchestration and management of the entire life cycle for network functions. The Cloud Platform software is decoupled from the Cloud Platform hardware (i.e. it can typically be sourced from different vendors).

There are many types of "O-Cloud" deployments.

- **Public cloud**: Shared, on-demand infrastructure and resources and generally a "Pay-As-you-Grow" model + third-party provider
  In this model, a third party (like Amazon, Microsoft, and Google) builds and manages the infrastructure that makes cloud computing possible. Service providers that do not want to be in the business of managing the cloud can access these shared computing resources and infrastructure on demand, either via the public internet or sometimes via private connections created by the provider.
- **Private cloud**: Dedicated, on-demand infrastructure and resources + owned data center
  The private cloud is where service providers operate their *own* infrastructure for cloud computing and accessed over a private network connection. Unlike public cloud, it is not shared with other organizations (unless its owner chooses to share). Many wireless service providers require a private cloud to tightly control, secure, and run their resources and infrastructure.
- **Hybrid cloud**: Public cloud + private cloud + consistent infrastructure and operations
  In a hybrid cloud model, service providers use a combination of public and private cloud resources on demand.
- **Multi-cloud**: Any combo of native public clouds and private clouds
  Some service providers combine public cloud and private cloud that meet the needs.
  This multi-cloud model provides ultimate flexibility to choose the cloud that best fits the application or business need (Figure 4.3).

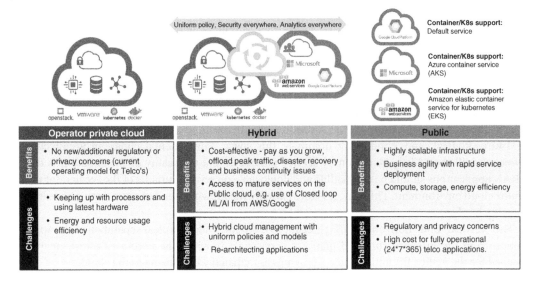

Figure 4.3 "Any Cloud" deployment models.

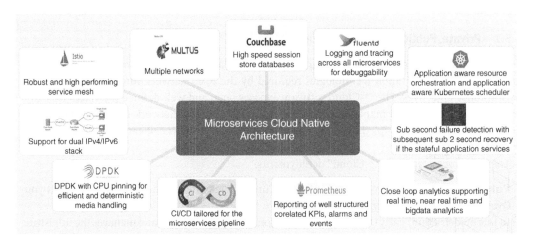

Figure 4.4 Features required for "Any Cloud" deployment.

4.3.2 Telco Features Required for "Any Cloud" Deployment

In terms of transport and networking challenges, following key items need to be considered for cloud deployments (Figure 4.4):

1) **MULTUS support**: MULTUS is a container network interface (CNI) plugin for K8S that enables attaching multiple network interfaces to pods. MULTUS allows pods to have multiple network interface connections that can address various use cases such as splitting control and data plane traffic. This needs to be supported in cloud platforms.

2) **VIP**: A VIP is an IP address that is assigned to multiple applications that reside on a single server, multiple domain names, or multiple servers, rather than being assigned to a specific single server or network interface card (NIC). The migration time to move IP address in a cloud deployment must be within a few seconds.

3) **SR-IOV support**: Single root I/O virtualization (SR-IOV) allows a device, such as a network adapter, to separate access to its resources among various PCIe hardware functions. This is required to be supported for RAN platforms in the cloud.
4) **DPDK support**: DPDK is the Data Plane Development Kit that consists of libraries to accelerate packet processing workloads running on a wide variety of CPU architectures. This needs to be supported on cloud platforms with high enough number of Resident set size (RSS) queues.
5) **Packet processing**: Throughput aspects in terms of packets processed per second must be considered.
6) **Network latency**: As cloud gets deployed with transport limitations, latency aspects need to be considered for various applications.
7) **IPv6**: As the number of public IPv4 addresses available can be limited, IPv6 support must be considered for cloud deployments.
8) **BGP support**: Border gateway protocol (BGP) is a standardized exterior gateway protocol designed to exchange routing and reachability information among autonomous systems (AS) on the internet. This is required for cloud deployments.

As we discuss cloud models, various questions get raised on the deployment challenges with the consideration for telco systems. Key areas of discussion involve:

a) **Maturity of cloud deployments**: The entire cloud model for telcos is just starting with the core networks moving to the cloud. The core is simpler to move to the cloud since it is more centralized and not sensitive to latency. As we go toward the RAN, which is deployed more at the edge and is decentralized with stringent latency constraints, the cloud model (esp. public cloud) validation has been debated. Some operators such as DISH Network have deployed RAN CU on amazon web services (AWS) cloud while the DU is still at a local data center or at a cell site.
b) **5-9s SLA achievable, with high availability regional cloud failure**: The expectation from telcos has been to provide the same service level agreements (SLAs) that they can currently achieve in a traditional deployment. There is concern that a public cloud failure can cause a wide outage, especially for critical emergency services such as E911. Hence, the redundancy and scalability aspects need to be carefully considered in deployments.
c) **Performance and cost benefits**: Telcos want to understand the tradeoffs between cost and performance for telco cloud deployments. While the initial pay-as-you-grow and have data is attractive, careful analysis needs to be done for ensuring over the long run, this remains financially attractive as well as enables performance to scale with deployments.
d) **Security**: Security has been one of the key concerns for cloud deployments – especially when public cloud is also included. Telco operators take a lot of effort to ensure network security is not compromised, and they want to have a better understanding of how this will work in a real deployment on the cloud before they decide to move in this direction.

4.3.3   On Premise, Far Edge, Edge, and Central Deployments

The decomposed RAN functions can be placed on premise (cell site)/far edge, edge, or regional cloud depending on the services and the required capabilities to offer the SLAs required for those services. In any implementation of logical network functionality, decisions need to be made regarding which logical functions are mapped to which Cloud Platform, and therefore which functions are to be colocated with other logical functions (Figure 4.5).

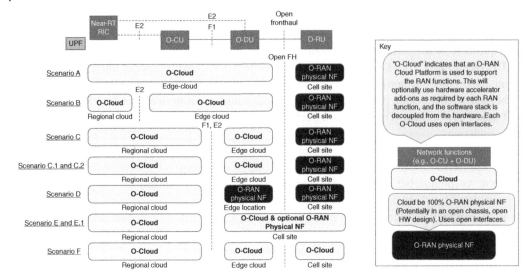

Figure 4.5 RAN function deployment. *Source:* O-RAN.

Service providers (communication service provider [CSPs]) and enterprise) adopt different cloud infrastructure options for different locations, i.e. edge could be public cloud while far edge may be private, and the nature of the cloud infrastructure provided across operators may vary. Therefore, it is critical that the applications are infrastructure agnostic. This also allows for independent upgradeability and scalability.

4.3.4 Classical, Virtual Machines (VMs), Containers, and Hybrid Deployments

While cloud "container" based deployments are rapidly evolving, brownfield deployments with a mix of classical, virtual machine (VM) based, container on bare metal, and container on VM combinations will continue to exist for many years to come (Figure 4.6).

4.4 Unwinding the RAN Monolith

RAN applications comprise of multiple functions that together perform specific tasks. In the pre-cloud era, these functions were grouped in a single monolithic code package. This type of architecture introduces many challenges in operation, management, and maintenance because the application needs to be deployed, scaled, and upgraded as a single entity. With the move to the cloud, a monolithic architecture is unable to take full advantage of cloud characteristics. The result is a transformation of the telephony concept of a network element to a network optimized distribution of functionality by adopting "cloud-native" architectural principles more consistent with the current industry trends and an automated, programmable, and intelligent orchestration and management of network functions is required to realize the advantages of cloud-based deployments (Figure 4.7).

The move to cloud-native RAN deployment necessitates a transformation in the decomposed RAN software architecture into a software designed, developed, and optimized to exploit cloud technology.

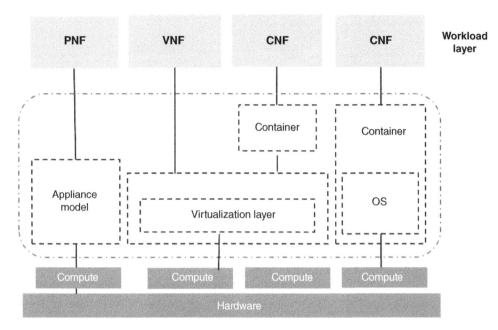

Figure 4.6 Workload deployment options.

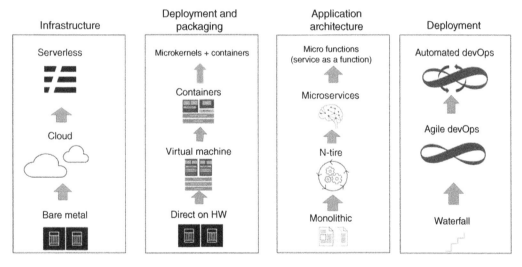

Figure 4.7 Aspects of transformation.

Microservices/containers, infrastructure agnostics, DevOps, and open and well-defined Application Programming Interfaces (APIs) are some of the key groupings of "cloud-native[1]" architectural tenets that have become critical in transforming how applications are developed, delivered, and optimized to exploit cloud technology (Figure 4.8).

---

1 "*Cloud-native technologies empower organizations to build and run scalable applications in modern, dynamic environments such as public, private, and hybrid clouds. Containers, service meshes, microservices, immutable infrastructure, and declarative APIs exemplify this approach. These techniques enable loosely coupled systems that are resilient, manageable, and observable. Combined with robust automation, they allow engineers to make high-impact changes frequently and predictably, with minimal toil.*" *Ref: cncf.io.*

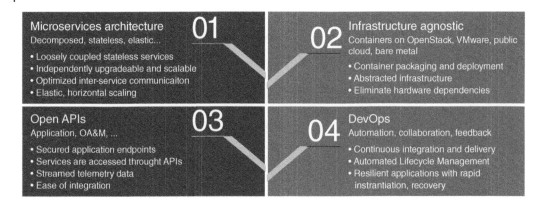

Figure 4.8 Cloud-native principles.

Building a cloud-native RAN necessitates a transformation in the RAN software architecture with software designed, developed, and optimized by applying cloud-native architectural tenets. RAN functionality is extremely sensitive to latency and throughput with stringent requirements and the "purist" view of cloud native influenced via Cloud Native Computing Foundation (CNCF) Telecom User Group (TUG), CNCF Cloud Native Network Functions (CNF) Working Group (WG), and Kubernetes needs to be adapted for *Cloud RAN*. Microservice architecture patterns influence the application's design and architecture. Even when most parts of the RAN functions (Operation, Administration, Management [OAM], CP, non-real-time functions) will be transformed to cloud native, some of the highly computation-oriented and latency-sensitive components will require adaptation.

Modular architecture is required to enhance cloud environment flexibility. However, the granularity of the breakdown must be architected carefully, otherwise it can lead to a degradation in performance and introduce latency as a result of increased overhead in the inter-communication between these modules. Therefore, a balance needs to be struck between granularity, performance, and ease of operation. It is generally a good practice to create modular software and based on functional affinity and interdependence, wrap them up in a single Pod to avoid inter Pod latency.

Cloud-native principles and microservices architecture patterns influence the application software design and architecture as we move to cloud-native RAN, and many of these principles can be applied both to classical and cloud deployments. Some key RAN characteristics are:

- Ultra-low latency
- Real-time handling
- High-precision timer/sync
- Specific HW acceleration card and CPU/GPU pinning for performance
- Ultra-high availability and redundancy

These types of requirements are necessary within the RAN and will govern the adaptation, feasibility, and complexity to achieve cloud native.

4.4.1 Adapting Cloud-Native Principles

Besides the obvious changes to adopt "cloud-native" principles in general, there are some points emphasized next that are typically complex to address with virtualized RAN, but critical to consider.

- **Clear separation of application business logic from infrastructure and platform**: The cloud infrastructure and platform ecosystem today vary widely across different service providers. Having an *abstraction layer* allows a clean separation of the application and the infrastructure/platform, enabling portability.
- **Support for deployment on various cloud infrastructures across multi-cloud and hybrid cloud support**: Dynamically instantiating network functions (NFs) at different points in the network demands that the application is portable across cloud ecosystem to accommodate service providers that deploy different cloud infrastructure at the edge, far edge, etc.
- **Standardized and automated container life cycle management and operability**: Standardized and automated container life cycle management, automonitoring, optimum placement of functions, and on-demand scaling of resources. This can be adopted with cloud and classical deployments. Kubernetes/Helm/European Telecommunications Standards Institute (ETSI) based for container orchestration are popular and many operators require it. There are several open-source initiatives like Nephio and XGVela that are providing concrete implementation of these solutions.
- **DevOps**: The scale and distributed nature of the RAN functions introduce challenges for distributing production software frequently. A cookie cutter approach to DevOps cannot be applied as the microservices architecture is tied to organization and process changes that can vary enormously based on the nature of the unique requirements of application and domain.

4.4.2 Architectural Patterns

- **Adopt techniques like multi-level modularity**: Decompose the software into the pieces that typically have low variance under load (e.g. parts of the scheduler that have pretty constant occupancy under widely varying user loads) and parts that are highly variant with load (e.g. per user processing). Engineer the highly variant parts into scalable modules to provide good tradeoffs in resource usage to actual load. Engineer the low variance pieces into nonscalable modules. This conserves development effort on scaling to those parts of the software that can benefit from it.
- **"Make before break" architecture design**: These architecture patterns are enablers to support software upgrade and resiliency.
- **Adopt a shared data layer with a high performance and reliable messaging system**: Specification and description language (SDL) enables sharing and monitoring of data via a generic API. SDL message exchange must be highly reliable, robust, secure, and fulfill required SLAs; SDL/network address translation (NATS) is an option to consider based on performance numbers.
- **Applying reactive architecture principles**: Reactive architecture is a set of design practices and architectural principles that application designers and developers apply to ensure that a distributed system can be responsive (serve data to its users in a timely fashion), resilient (always available, self-healing), and elastic by ensuring efficient management of distributed state and communication. Reactive architecture design patterns are useful for intelligent forms of data replication, coordination, and persistence and use different flavors of message-driven architectures, such as Command Query Responsibility Segregation (CQRS) and event sourcing, to treat a distributed system's state.

- **Leverage service mesh[2] for handling service-to-service communication**: Adopting microservices architecture results in complex service and application topologies as the number of components increases. This leads to new set of problems in cloud-native deployments: service discovery, traffic management, security and policy management, and observability. The problems can be solved by either making the problem a part of the application itself, where additional logic is coded into the business logic of the application, or, bringing in a separate infrastructure solution like service mesh.

- **A balance between standards and open ecosystem**: 3GPP has driven a service-based architecture for the 5G core, while the 5G RAN is still very broadly categorized into RU, DU, CU-CP, and CU-UP as functional components with very transactional interfaces (3GPP) defined between them. 3GPP interfaces are sticky and many have been adopted in O-RAN. These interfaces have to be adapted to enable cloud nativity. These functions, as a monolith, containerized, have some degree of scalability via growth procedures, but this is often non-trivial and requires complex engineering and services efforts and scaling the capacity of the network often involves adding new NFs as well as configuring connectivity between the new NFs.

   Interfacing and embracing *open-source* software for applications, platform, and cloud orchestration services are necessary for faster time to market and greater interoperability and ease of deployment across different ecosystems in customer environments. Open-source components are playing a very critical role in this transition for telcos to cloud native.

4.4.3 Software Architecture Portability and Refactoring Considerations

- **Software base portable across legacy hardware, bare metal, public cloud**: Many service providers will continue with classical RAN deployment.

- **Careful consideration of inter Pod/VM communication performance penalty when splitting latency/reliability-sensitive functions across Pods**: Inter Pod communication should be avoided for latency-sensitive functions. Techniques like shared memory communication first between containers, then between Pod's, and ultimately between nodes via fast memory interconnect are expected to give us more speed-up for internal communication. For security reasons, service providers require inter Pod communication to be encrypted. This will add an additional tax on latency and affect performance. Techniques like Pod affinity can be used to mitigate this. Inter Pod communication reliability needs to be considered, as well, when splitting functions. Message coding (like ASN.1) needed for backward compatibility support for software (SW) updates cause high latencies and there is need to use more novel methods for that purpose.

- **Stateless and stateful or state-efficient**: While stateless computations are necessary from a cloud-native "purist" perspective, they can incur significant overhead, even when the data and the computation are on the same machine. To support latency-sensitive, real-time applications in Layer 1, scheduler applications, Layer 2 real-time RAN applications, we have to adapt the definition of "stateless." Colocating data and supporting state-efficient computations (encapsulate mutable states, asynchronous, and non-blocking execution with reactive architecture design patterns) enable avoiding costly state transfers between successive operations. The "state-efficient" concept is also sometimes referred to as "state-optimized" or "state-cached."

---

2 A **service mesh** is a dedicated infrastructure layer for handling service-to-service communication for ensuring reliability, security, intelligent routing, and traceability of communication among the services that comprise a cloud-native application. *References: https://medium.com/microservices-in-practice/service-mesh-for-microservices-2953109a3c9a "Service Mesh for Microservices," Kasun Indrasiri, http://Medium.com, 15 September 2017.*

4.4.4 Compute Architecture Consideration

- **Clean abstraction layer** to enable portability across a multitude of hardware platforms (CPU, FPGA, SoC, GPU, accelerators, etc.):
  The hardware platform ecosystem today varies widely, and the RAN Layer 1 software and firmware are tightly coupled. Having an *abstraction layer* allows a clean separation of the application and the infrastructure, enabling *portability*.
- **Inline/look aside architecture performance consideration**:
  While "look-aside" architecture may be modular, the "inline" model (Figure 4.9) provides improvements for the real-time performance. However, with the abstraction layer there should be flexibility to discover the capabilities of the accelerator and determine what approach can be used.[3]

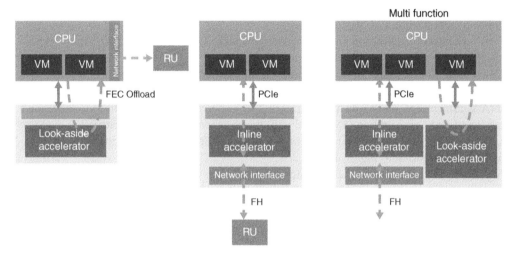

Figure 4.9 In-line and look-aside accelerators.

- **Real-time task scheduling and core pinning**
  Resource sharing is enabled by a real-time task scheduler, which allows multiple tasks to be load balanced across multiple cores. This is done by a low overhead, non-preemptive task dispatcher polling tasks from multiple queues with each task running to completion. Operating system (OS), infrastructure, and other non-real-time tasks are given separate resource allocations more typical to cloud native.

  In typical cloud-native non-real-time applications that tolerate variance in processing deadlines, cores/resources are shared, and not pinned. Resource sharing is usually enabled by OS time slicing allowing for efficient use of underlying resources (i.e. load balancing and pooling). Real-time RAN cloud applications are also shared and sometimes pinned. These ultra-real-time applications have strict deadlines (e.g. on the order of 1 us) and therefore only tolerate minimal processing variance. An OS-based sharing mechanism cannot meet such constraint. Pinning enables a process to run on a dedicated core and that way helps to meet short duration critical deadlines for these ultra-real-time applications. Kubernetes enables dedicated resource allocation through use of resource management plugins like a CPU manager.

---

3 Hardware acceleration – Wikipedia.

By enabling hyperthreading, one may get additional performance out of the host. Hyperthreading, however, can increase the variance of real-time processing. Hyperthreading gains are typically quoted based on same application sharing sibling threads. This creates additional constraints on the pinning strategies.

4.5 Orchestration, Management, and Automation as Key to Success

The decoupling of RAN hardware and software for all components including O-CU (O-CU-CP and O-CU-UP), O-DU, and O-RU and the deployment of software components on commodity server architectures supplemented with programmable accelerators where necessary have been key drivers in moving to the cloud. A 5G network with disaggregated RAN applications, disaggregation of hardware and software, and (multi) cloud-based deployments will require intelligent and programmable orchestration and management.

O-RAN architecture definition and mechanism introduced for orchestration and management will facilitate ease of cloud-based deployment (Figure 4.10).

O-RAN WG6 Cloudification and Orchestration group currently has two main working subgroups: Cloud architecture and O2 interface for life cycle management (LCM) and acceleration abstraction layer (AAL) subgroups. The hardware components are exposed as logical functions to the Service Management and Orchestration (SMO) and with the combination of the infrastructure management function and deployment management functions, the SMO orchestrates the life cycle management of the application (Figure 4.11).

There are two main interfaces that are being defined:

- LCM of the cloud infrastructure and orchestration/instantiation of different types of applications (RAN, non-RAN, etc.) via the O2 interface.
- Provide fine grain control of the acceleration functions and task handling via the AALI interface.

The aim is to:

- Enable flexible instantiation and life cycle management through orchestration automation.

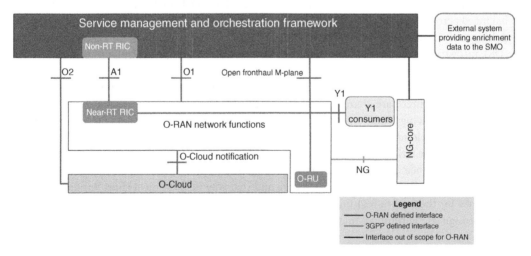

Figure 4.10 Service management and orchestration. *Source:* O-RAN.

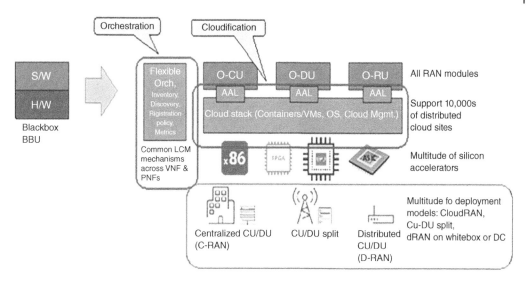

Figure 4.11 Infrastructure and deployment management services. *Source:* O-RAN.

- Promote the availability of a multitude of hardware implementation choices for a given software implementation.
- Enable workloads to move across clouds while leveraging the hardware variant available in that cloud resource pool with a central orchestrator framework.
- Enable elastically scalable capacity for on-demand and as services evolve.
- Enable rapidly deployed services – minutes instead of days, weeks, or months.
- Enable decoupling of the hardware from the software requiring intelligent and flexible instantiation and life cycle management of the software.
- Enable sharing open platform services (PaaS) across applications through common vendor-neutral mechanisms/APIs for hardware discovery/deployment/management, life cycle management, and orchestration of RAN and non-RAN applications.
- Enable setting up autonomous notifications as well as on-demand query of the inventory and the capabilities.
- Enable allocating/selecting compute resources during instantiation of the services (e.g. vDU resources pooled or dedicated per slice) and adaptably deploy applications (RAN, non-RAN).
- Enable monitoring the hardware resources (telemetry, status, etc.).
- Support a generalized hardware abstraction layer to hide the lower layer details of the hardware from the application, enabling portability.

The Cloud resources include compute, networking, and storage components and may also include various acceleration technologies required by the RAN network functions to meet their performance objectives. O-RAN is in the process of defining a logical O2 interface between the SMO and O-Cloud for discovery of cloud resources, LCM, and orchestrations of applications in a multi-cloud, multi-vendor environment. It also enables the management of O-Cloud infrastructures and deployment of an O-Cloud (Figure 4.12).

- **Infrastructure Management Services (IMS)**: The IMS are responsible for management of the O-Cloud resources and the software that is used to manage those resources. The IMS generally provide services for consumption by the FOCOM.

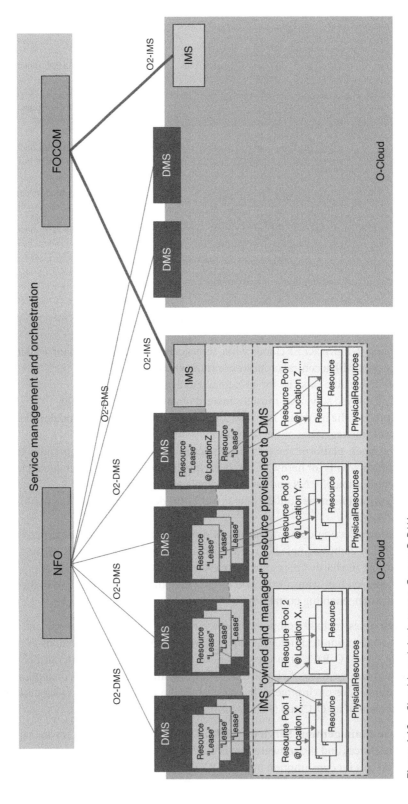

Figure 4.12 Cloud-based deployment. *Source:* O-RAN.

- **Deployment Management Services (DMS)**: The DMS are responsible for management of NF deployments into the O-Cloud. They provide the ability to instantiate, monitor, and terminate NF deployments. The DMS generally provide services for consumption by the NFO.
- **Federated O-Cloud Orchestration and Management (FOCOM)**: The FOCOM is responsible for accounting and asset management of the resources in the cloud. The FOCOM is the primary consumer of services provided by the IMS. The FOCOM has information about the O-Cloud resources management. Specifically, the FOCOM needs to know whether the services are within the operator domain or external.
- **Network Function Orchestrator (NFO)**: The NFO is responsible for orchestrating the assembly of the network functions as a composition of NF deployments in the O-Cloud. It may also utilize OAM functions in order to access the O1 interface to the NF once it is deployed. Its use of the O1 is not germane to the O2 and is only mentioned here for completeness. The NFO is the primary consumer of the DMS.
- **OAM functions**: The OAM functions are responsible for Fault, Configuration, Administration, Provisioning, Security (FCAPS) management of O-RAN managed entities. In the context of O2 they may be one of the functional blocks for which a callback is provided in order to receive faults and/or performance data from a subscription for O2ims or O2dms procedures

There are several existing and de facto standards that are being leveraged to accomplish these goals; for example, K8s, ETSI, 3GPP, and open-source initiatives like DPDK, ONAP, and various Linux foundation community projects like XG Vela, CNCF, and Anuket will be leveraged.

The challenge, however, is the proliferation of these different standards, and open-source activities that will create too many variations, sometimes complementary but most of the time divergent, functional specifications resulting in an interoperability nightmare. There is conscious effort being made in O-RAN to reuse and leverage existing definitions.

4.5.1 Acceleration Abstraction Layer

Hardware acceleration is the use of specially developed computing hardware to perform some functions more efficiently than it is possible in software running on a general-purpose CPU. The ability to dynamically offload compute intensive, real-time processing to the accelerator is critical to support the services envisioned for 5G (ultra-low latency services, AI/ML algorithms, crypto, etc.) (Figure 4.13). Note: Figure 4.13 is an adaptation based on the O-RAN specification.

Hardware accelerators from different vendors can have different capabilities. Lack of the common abstraction to hide the diversity of these hardware variants makes applications portability difficult. In addition, dynamic service orchestration, resources allocation, and programmability for fine grain control require open and standardized interfaces. The abstraction layer will provide a standard interface and set of software/firmware services for installing and communicating between applications and hardware acceleration modules over a standard interface (e.g. PCIe) with the ability to:

- Support a generalized hardware abstraction layer to hide the lower layer details of the hardware from the application enabling portability.
- Partition the accelerator and expose it as a logical processing unit (AAL-LPU).
- Monitor the hardware resources (telemetry, status, etc.).

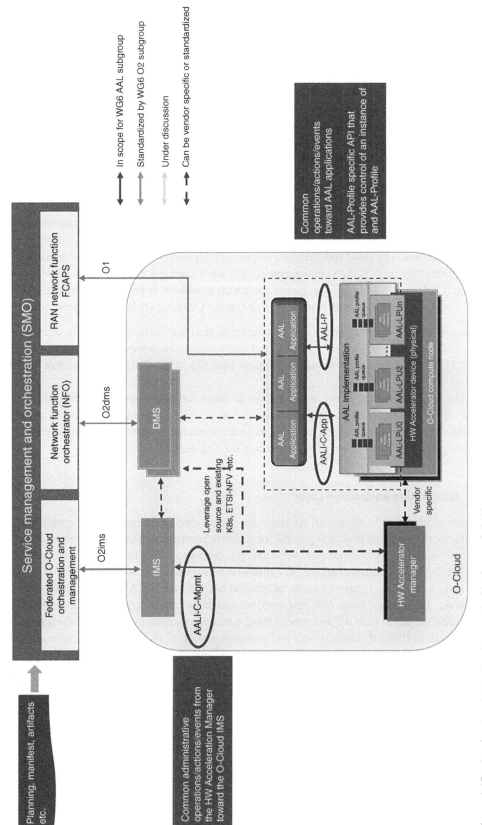

Figure 4.13 Accelerator abstraction layer and APIs. *Source:* O-RAN.

- Allocate/select compute resources during instantiation of the services (e.g. vDU resources pooled or dedicated per slice).
- Fine grain control from the application to "learn" capabilities of the hardware and manage/ optimize accordingly.

O-RAN is in the process of specifying AAL APIs for abstracting the low-level details of the hardware accelerator technology.

The AAL interface has two distinct parts: the first part corresponds to a set of common interfaces (AALI-C). These are used to address all the profile-independent aspects between application(s) and underlying AAL implementation(s) including management and orchestration within an O-Cloud platform.

Within the AALI-C interface there are two distinct buckets:

AAL-C-Mgmt: Common administrative operations/actions/events toward O-Cloud IMS. The HW accelerator manager terminates the AALI-C-Mgmt interface.

AAL-C-App: Common operations/actions/events toward RAN application.

While there are many questions on the merit of compromising performance over portability of L1 RAN application, there are no doubts about the benefits of an AALI Common API for management that enables ease of standing up a workload and assigning accelerator resources to it.

4.5.2 Cloud Deployment Workflow Automation

Incumbent service providers transitioning to 5G have to navigate all sorts of constraints. Somehow, the groundbreaking new capabilities they are building need to fit within pre-existing network architectures, operations frameworks, cell tower footprints, and more. But when you are new to the mobile entrant, building a brand-new 5G network from scratch, the challenges are entirely different.

Cloud-based deployment requires coordination across different layers of the horizontal disaggregated stack and the different layers of vertical disaggregated stack. Leveraging open-source tools and automation of the workflow engine is necessary to ease orchestration across infrastructure resources, platform resources, and standup application workloads.

Continuous delivery/deployment of cloud-native network functions introduces the challenge of managing a network where its operability meta-models are continuously changing: every time that a new cloud-native network function release is deployed to the network there is the potential that its operability meta-model is changed:

- New metrics, alarms, and attributes might be introduced, and existing metrics, alarms, and attributes might be changed.
- The overall representation (managed object classes, hierarchy, and grouping) might be changed.

All those changes need to be managed by the network management without any changes to its own release cycle, i.e. there must be a total decoupling between cloud-native network function releases and network management releases.

Network management must support introduction of cloud-native network function operability meta-models during runtime, and it also needs to be able to cope with different networks that will have a different interpretation of the composition of a cloud-native network function; whereas in traditional networks the model composition (managed object classes, hierarchy, grouping) is done during development time, in cloud-native networks the model composition is done during runtime.

4.6 Summary

Cloud deployment strategy is generally to "Build Once, Deploy and Manage on Any Cloud" and more and more customers are choosing multi-cloud cloud options and associated platform services. The decoupling of RAN hardware and software for all components including O-CU (O-CU-CP and O-CU-UP), O-DU, and O-RU and the deployment of software components on commodity server architectures supplemented with programmable accelerators where necessary have been a key driver in moving to the cloud. The vertical and horizontal disaggregation requires intelligent and flexible orchestration and life cycle management of the O-Cloud infrastructure and application software. Automation will drive success and help operators realize the full potential of cloud-based deployments.

In order to achieve these goals architectural transformation and adoption of cloud-native principles are necessary. However, as highlighted in this chapter, cloud-native principles have to be adapted to ensure stringent performance criteria for real-time application software like the vDU.

Pure cloud-native networks will take time to realize; this means that there will be a significant amount of time where hybrid networks (a mix of physical, virtual, cloud, and cloud-native networks) will be prevalent in most operators.

There are several standards organizations, fora, open-source projects, and de facto standards that are being leveraged to achieve these goals. The O-RAN-defined O2 and the AALI interfaces enable applications to utilize the underlying infrastructure and platform services minimizing custom implementation.

Bibliography

O-RAN WG6 specification, v4.00 July 2022: O-RAN Cloud Architecture and Deployment Scenarios for O-RAN Virtualized RAN.

O-RAN WG6 specification, v6.00 November 2022: O-RAN Cloudification and Orchestration Use Cases and Requirements for O-RAN Virtualized RAN.

O-RAN WG6 specification, v3.00 November 2022: O-Cloud IMS and DMS Relationships to SMO: O2 Interface General Aspects and Principles.

O-RAN WG6 specification, v3.00 November 2022: O-RAN Acceleration Abstraction Layer Common API.

5

RAN Intelligence

Dhruv Gupta[1], Rajarajan Sivaraj[2], and Rittwik Jana[3]

[1] AT&T, San Ramon, CA, USA
[2] Mavenir, Richardson, TX, USA
[3] Google, New York City, NY, USA

5.1 Introduction

Mobile networks across the globe are poised to benefit from three key technology areas: advances in software-defined networking (SDN), evolution of cloud-based radio access networks (RANs), and rapid progress in the fields of artificial intelligence (AI) and machine learning (ML). These technologies offer tremendous new opportunities for operators not only by enabling innovative services and new revenue streams but also by achieving significant cost reduction by applying intelligent automation in all aspects of a mobile network. Intelligence across the network, and within individual network functions, is expected to drive innovative solutions for network planning, engineering, and operation.

There are three key enablers required to bring such an intelligent and autonomous network to fruition – the ability to monitor the network with the right dataset available in a timely manner, a rich ecosystem of artificial intelligence (AI)/machine learning (ML)-driven applications to optimize and heal the network, and last, the ability to control and guide the network based on operator's business requirements. O-RAN-defined interfaces (O1, A1, and E2) and controller platforms (non-RT and near-RT RIC) aim to enable this vision. (See O-RAN.WG1.O-RAN-Architecture-Description-v06.00 2022; O-RAN.WG2.A1TS-v02.00 2022; O-RAN.WG3.E2TS.V01.00 2022.)

5.2 Challenges and Opportunities in Building Intelligent Networks

Traditional radio resource management (RRM) solutions, largely based on heuristics, do not sufficiently account for intricacies resulting from rapidly changing wireless network dynamics. They are not adequately optimized to handle user- and service-centric optimization decisions for key RAN functionalities (such as connected and idle mode mobility, radio bearer admission, radio resource control and spectrum allocation, multi-radio access technology (multi-RAT) dual connectivity, carrier aggregation, and dynamic spectrum sharing) pertaining to evolving use cases and slicing requirements for 5G and beyond. This necessitates the need to have more data-driven, AI/ML-based solutions that can *learn* intricate interdependencies between RAN parameters, arising from complex interactions across the layers of the RAN protocol stack due to RRM decisions, and quantify their impact on individual user equipments (UEs) and collectively on the entire network.

Taking the example of the traffic steering feature to control the mobility of UEs in RAN, handover optimization is an age-old problem in cellular RAN, solutions for which have been widely studied and implemented. However, the requirements and deployment scenarios keep changing with evolving radio access technologies (RATs), newer use cases, and slicing requirements that traditional handover procedures and optimization techniques are not designed to handle. To illustrate this further, even as handover processing has been featured in third generation partnership project (3GPP) specifications for a while now since the 2G days (see 3GPP 2017), 3GPP standards for 5G, as recent as Release 16 (2020) (see 3GPP (2021)), have introduced a new handover feature, called dual active protocol stack (DAPS) handover (which enables the UE to stay connected to the same serving cell, even after receiving the handover command from the O-CU-CP hosting the cell up until the UE establishes a successful random access channel (RACH) to the target cell, thereby avoiding interruption in connectivity and data transfer). This is beneficial for processing the handover of ultra-reliable low-latency communications (URLLC) UEs that come with stringent latency control-plane and user-plane requirements. Likewise, traditional RRM or legacy self-organizing network (SON) solutions for handover, largely based on heuristics involving signaling measurement and load thresholds for cells, are not primed to handle optimal UE-centric handover decisions for serving new use cases and slicing requirements.

Toward this end, the near-RT RIC (near real-time RAN intelligent controller) leverages fine-grained UE-level intelligence, consisting of UE-specific performance measurements (PMs), key performance indicators (KPIs), UE-specific state variables across the layers of the RAN protocol stack, UE-specific parameters exchanged across network interface procedures, information elements pertaining to entities of the individual UEs, such as data radio bearers (DRBs), quality of service (QoS) flows, protocol data unit (PDU) sessions, physical resource blocks (PRBs) and logical channels pertaining to the UEs, toward making optimized RRM decisions, down to the granularity of individual UEs, for a plethora of O-RAN use cases. For example, various traffic steering strategies can be employed (e.g. finding the optimal primary cell of the UEs, choosing the optimal selection/reselection of secondary cells for the UE in carrier aggregation, optimizing cell reselection priority for the UEs, or optimizing secondary node selection in multi-RAT dual connectivity). Another example is Quality of Experience (QoE), which optimizes the QoS profile of the DRBs and optimally multiplexes the QoS flows to DRBs. Slicing assigns the PRB allocation quotas per-UE per-slice and optimizes the multiplexing of slice-specific PDU sessions to DRBs. Finally, for massive multiple-input multiple-output (MIMO) optimizes the number of layers and selects the optimal transmission mode, optimal synchronization signal block (SSB), and channel state information-reference signal (CSI-RS) beamforming weights, to name a few. The near-RT RIC is also guided by AI/ML-driven declarative policies and enrichment information, generated from the non-RT RIC via the A1 interface.

5.3 Background on Machine Learning Life Cycle Management

The ML deployment scenario, as detailed in O-RAN AI/ML technical report, is presented in the diagram shown in Figure 5.1. In general, the following steps are followed (not necessarily in strict ordering) and the role of AI/ML in the near-RT RIC is discussed here:

- The data for offline training from the E2 nodes (O-CU-CP, O-CU-UP, O-DU, O-eNB), the near-RT RIC, and the O-RU is sent over the O1 and open M-plane fronthaul interfaces to

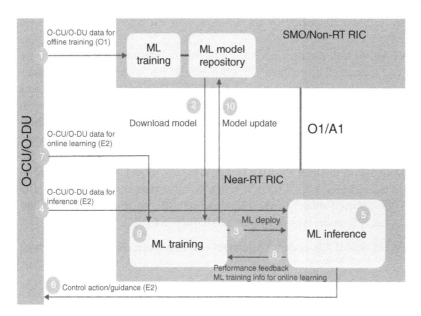

Figure 5.1 Machine learning deployment scenario. *Source:* O-RAN.WG2.AIML-v01.03, Oct. 2021

the service management and orchestration (SMO). This data includes (but is not limited to) PM data, KPI data, configuration management (CM) data, fault management (FM) – alarms and threshold crossing data, network interface trace data or UE-specific minimization of drive test (MDT) trace data, etc. This data is collected and stored in the SMO/non-RT RIC framework.

Previously, a PM/FM/trace job control activation would be initiated from the SMO to the E2 node and/or O-RU functions, either via some preconfigured application in the SMO/non-RT RIC framework or via the rApps deployed in the non-RT RIC.

- Upon data collection, the data is prepared, aggregated, and fed to the AI/ML training host in the non-RT RIC/SMO framework toward building the AI/ML training models, based on the choice of AI/ML algorithms (such as reinforcement learning (RL) – Q-learning/DQN, recurrent neural networks – long short-term memory (LSTM), Autoregressive Integrated Moving Average (ARIMA) time series prediction, deep learning, and supervised learning techniques). The AI/ML training host takes the aggregated data (PM/CM/FM/trace/topology, etc.) and the parameters to be optimized (target cell for HO, MU-MIMO layers, etc.) and trains the ML model based on a given choice of AI/ML algorithm towards meeting the target goal (maximize throughput, minimize latency, maximize reliability, minimize call drop, etc.). Previously, the SMO/non-RT RIC would have received a request from the near-RT RIC for offering AI/ML training services.
- The trained ML model is saved in the ML repository in the SMO/non-RT RIC framework and is published to the ML catalog maintained in the SMO/non-RT RIC framework. Once the non-RT RIC responds to the near-RT RIC via the A1 interface on the availability of the ML model sought by the near-RT RIC along with details of the ML model catalog and the information on the repository where the ML model is stored, the ML model can be downloaded via the O1 interface and deployed in the inference engine of the near-RT RIC.
- The near-RT RIC can receive data from the E2 nodes (O-CU-CP, O-CU-UP, O-DU, O-eNB) via the E2 interface into the inference engine where the ML model is deployed. This data is based on

the E2 service models (E2SMs) supported by the E2 nodes for the near-RT RIC use cases and xApps.

- Based on the deployed ML model, the near-RT RIC can perform inference based on E2 data reports received via E2AP indication procedures. Once the near-RT RIC makes inference, it generates control action or imperative policy guidance via E2AP CONTROL or POLICY procedures back to the E2 nodes via the E2 interface using the respective E2SMs.
- The deployed ML model can also be subject to online ML training updates based on the subsequent E2 data received by the near-RT RIC from the E2 nodes, and based on continuous monitoring of network performance data and the feedback generated by the ML inference engine to the online ML training engine in the near-RT RIC. The updated ML model is again deployed in the inference engine of the near-RT RIC. If there is a significant update to the ML model, the near-RT RIC uploads the updated model to the SMO/non-RT RIC via the O1 interface and the model is updated in the ML model repository in the SMO/non-RT RIC.

These steps are captured in the ML life cycle management and implementation example shown in Figure 5.2.

5.4 ML-Driven Intelligence and Analytics for Non-RT RIC

The life cycle of RANs involves two primary stages: the network planning/deployment phase and the network operations/assurance phase. The network deployment phase involves the planning, installation, and configuration of RAN network functions/elements. A typical process may involve the operators starting with a set of design targets defined at a system level or predefined user-group level. Based on these, the operator will preconfigure a variety of attributes in the network functions/ elements in order to meet the defined design targets (for example, radio frequency (RF) power, frequency bands, QoS attributes, and so on). This determines the default behavior of the network in terms of satisfying the business requirements and SLAs.

The process of RAN service and performance assurance involves a feedback loop consisting of continuous monitoring of the network, followed by adjustment of default/existing network configuration to achieve design targets under dynamic changes (for example, network load, topology, new services, etc.). Traditionally, operators have achieved this via a combination of manual engineering and cSON (centralized SON)-based automated control loops.

However, with the onset of 5G technology, it has become clear that these approaches need to evolve, in order to enable more customer-centric control and allow for rapid optimization at a larger scale. Legacy solutions that are based on system- or node-level configuration do not always help in meeting the desired optimization objectives. Specifically, while such solutions are apt for handling network-wide objectives around evenly distributing traffic load or energy optimization, they are not as useful in other areas such as providing service-layer service level agreements (SLAs) to end customers. Such use cases are increasingly important for operators from the perspective of supporting new and improved 5G services that can help in revenue generation. In order for operators to be able to provide differentiated services to customers (for example, high throughput, low latency, etc.), it is imperative that operators can perform network assurance at an individual customer level.

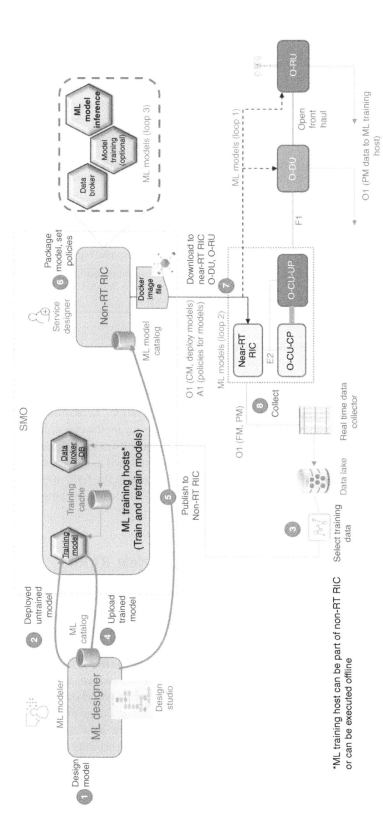

Figure 5.2 Machine learning life cycle management and implementation example. *Source:* O-RAN.WG2.AIML-v01.03, Oct. 2021

The following table compares the legacy or present mode of operation for SON and automation and the future direction being led by the open RAN movement.

Category	Legacy approach (SON)	Future "Open RAN" approach
Scope	Separate RAN, core, and transport-focused solutions and systems	Combined view and control of mobility network
Granularity	Coarse node-/system-level configuration parameters	Finer granularity of control including subscriber level and slice level
Openness	Reliant on vendor proprietary APIs and data models	Based on open standards such as O-RAN A1 and O1 interfaces
Intelligence	Decisions driven by aggregated node-level counters, limited use of advanced AI/ML	Autonomous AI/ML-driven decision-making based on fine-grained and timely data, including external enrichment data
Impact	Restricted to a static set of configuration parameters, the impact is not deterministic	Combination of configuration and dynamic policies; deterministic outcome of control actions

Based on these objectives, O-RAN Alliance has defined the non-real-time RAN intelligent controller (non-RT RIC) whose main objective is to drive RAN resource optimization in an intelligent manner while leveraging open standardized interfaces. For this purpose, the non-RT RIC can leverage the O1 interface and the A1 interface. The O1 interface provides the traditional Operations, Administration and Maintenance/Fault, Configuration, Accounting, Performance, Security (OAM/FCAPS) functionality, including providing the key inputs to non-RT RIC in terms of RAN PM data and events, and transmitting the outputs in the form of provisioning or configuration parameters. The A1 interface, on the other hand, was introduced to allow the non-RT RIC to provide other forms of external inputs into the RAN. This can take the form of policies that can help guide the RAN functions toward a desired objective, or in the form of Enrichment Information, which can serve to enhance the RAN nodes' decision making.

5.5 ML-Driven Intelligence and Analytics for Near-RT RIC

The near-RT RIC is responsible for exercising fine-grained RRM decisions for RAN functionalities and related parameters, largely pertaining to the control-plane (C-plane) and user-plane (U-plane) procedures associated with the layers of the RAN protocol stack, associated with individual UEs and layer entities, such as DRBs, QoS flows, slices, and IP PDU sessions. Toward this end, the near-RT RIC shall be guided by policies and KPI target objectives set by the non-RT RIC (see Chapter 3). This distinguishes the near-RT RIC from legacy SON engines, which focus on optimizing management-plane (M-plane) operations of network elements and associated cell-level parameters. SON implementations today have major interoperability issues due to multi-vendor environment and configuration inconsistencies. Integrating multi-party SON solutions in HetNet deployments leads to the degradation of overall KPIs. There is an opportunity to rethink how SON/RRM use cases can be tackled more holistically. Specifically, as one option, the SON function realization can be strictly restricted to the O-RAN architecture, i.e. implemented either in the near-RT RIC or the non-RT RIC (this includes the "SON coordination function"); another option could be RRM functionalities like Call Admission Control (CAC), Mobility Load Balancing (MLB), Mobility Robustness Optimization (MRO), etc., to be implemented at the E2 node itself with a possible

"RRM Optimization xAPP" at the near-RT RIC. Apart from these, there are other aspects of distinction between the near-RT RIC and SON engines.

- **Intelligence:** While SON engines broadly apply cell-level analytics and data correlation from performance metrics generated from across the RAN protocol stack layers managed by the network elements, the near-RT RIC additionally performs UE-level cross-layer analytics and data correlations of UE-specific state variables and contextual information obtained from across the RAN protocol stack.
- **Interfaces:** While SON engines use proprietary interfaces to manage the network elements, the near-RT RIC employs O-RAN-standardized open interfaces for fine-grained RRM of UEs, cells, slices, and E2 nodes, thereby achieving multi-vendor interoperability.
- **Granularity:** While SON engines control parameters concerning the M-plane of the network elements in non-real-time (>1 s) at coarser granularities, the near-RT RIC performs fine-grained control of C-plane and U-plane parameters associated with UEs and RAN protocol stack entities in near-real-time (10 ms to 1 s) using low-latency control loops.
- **Policies:** Prebuilt policies are deployed on SON engines to operate on network data toward controlling M-plane parameters. These policies may be trained offline using ML models and later deployed in SON engines. The non-RT RIC and near-RT RIC shall, on the other hand, adaptively update fine-grained policies online using O-RAN-defined ML life cycle management based on continuous monitoring of network performance over standard interfaces.

The customized UE-level decisions at finer granularities, backed by programmable and adaptive intelligence, are the centerpiece of the RIC, which distinguishes itself from SON. The reasons why customized UE-level decisions are important are due to the rapid proliferation of fifth-generation (5G) telecommunication services with strict QoS/QoE demands that are expected to generate an exponential increase in mobile data traffic and promise to yield an unprecedented user experience. With the evolution of telecommunication services, 5G telecommunication services focus on the following categories of use cases, namely:

(i) **Enhanced Mobile Broadband (eMBB):** for bandwidth-intensive HD 4–8 K video/VR streaming, immersive augmented reality (AR)/virtual reality (VR), etc.
(ii) **Ultra-Reliable Low-Latency Communication (uRLLC):** for intelligent transportation services, factory automation, remote telesurgery, real-time drone surveillance, etc.
(iii) **Massive IoT (MIoT):** for wearables, etc., requiring high coverage to support network densification.
(iv) **High-Performance Machine-Type Communication (HMTC):** for mission-critical communication, requiring ultra-high reliability and high availability/coverage of the network.
(v) **V2X (Vehicle to Everything):** for intelligent transportation services, connected vehicles, autopilots, self-driving cars, etc.

Beyond 5G networks shall focus on provisioning newer use cases like metaverse, telepresence, etc., that shall deliver an altogether new experience to mobile UEs. Enterprises expect mobile network operators to deliver such applications to UEs with QoS assurances in network performance that enrich the end UE's QoE. Toward this end, mobile network operators are *slicing* their network resources for serving these use cases.

5.6 E2 Service Models for Near-RT RIC

The E2SM describes the functions in the E2 node, which may be controlled by the near-RT RIC and the related procedures, thus defining a function-specific RRM split between the E2 node and the near-RT RIC. They describe a set of services exposed by the E2 node that shall be subsequently used by the near-RT RIC and the hosted xApps. These services provide the near-RT RIC with access to messages and measurements exposed from the E2 node (such as cell configuration information, supported slices, public land mobile network (PLMN) identity, network measurements, and UE context information) that enable control of the E2 node from the near-RT RIC. Multiple E2SMs have been defined in O-RAN WG3 such as E2SM-RAN Control (E2SM-RC), E2SM-Key Performance Monitoring (E2SM-KPM), E2SM-Network Interface (E2SM-NI), and E2SM-Cell Configuration and Control (E2SM-CCC). In this section, we discuss how E2SMs are used in building ML models for xApps in the near-RT RIC. We further discuss how E2SMs are used in managing the ML life cycle and continuous performance monitoring, involving the near-RT RIC.

5.6.1 E2SM-KPM

Using E2SM-KPM, the E2 node can stream UE-level, cell-level, and E2 node-level PM data across the layers of the RAN protocol stack at near-real-time granularities (ranging from 10 ms to 1 s) to the near-RT RIC.

The PMs, standardized in O-RAN WG3, include packet delay measurements at Packet Data Convergence Protocol (PDCP), Radio Link Control (RLC), Medium Access Control/Physical (MAC/PHY) layers, radio resource utilization measurements, UE throughput measurements, radio resource control (RRC) connection number/connection establishment/re-establishment measurements, mobility management measurements (number of intra-RAT and inter-RAT handovers), transport block (TB)-related measurements (number of TBs modulated with quadrature phase shift keying (QPSK), 16QAM, 64QAM, 256QAM), CQI-related measurements (wideband and sub-band CQI), QoS flow-related measurements (number of QoS flows setup, release, and modification), DRB-related measurements (number of DRBs setup, release, attempts, and in-session activity time), received random access preambles per cell or SSB, distribution of RSRP values per SSB, number of UEs with active DRB transmission, packet loss rate due to over-the-air transmission losses, packet drop rate due to high PDCP traffic load, PDCP data volume measurements in terms of amount of successfully transmitted PDCP SDU bytes, IP latency measurements due to buffering in the RLC layer caused by network congestion, UE and bearer context release measurements (average and distribution), call duration, the average or distribution of RSRP/RSRQ of UEs subject to handover with respect to the serving cell and the target cell, etc.

E2SM-KPM enables streaming these measurements from the E2 node to the near-RT RIC at periodic intervals. It is to be noted that while 3GPP TS 28.552 and TS 32.425 discuss streaming these PMs at a cell level or an E2 node level, E2SM-KPM additionally facilitates the reporting of these PMs at a per-UE level at near-RT periodicities.

5.6.2 E2SM-RC

Using E2SM-RC, the E2 node can stream or send the following information at UE-level to the near-RT RIC at near-real-time periodicities, such as:

1. Context information
   UE-specific RAN state and context information (L2 PDCP/RLC/MAC state variables, RLC buffer occupancy, etc.), E2 node information (serving cell context information, neighbor cell information, etc.), UE-specific L3 RRC measurements (serving cell RRC and neighbor cell RRC measurements), and RRC state information of the UEs.

2. UE-specific signaling information
   This includes information about the slice profile and the PDU sessions subscribed by the UEs, information about the QoS flows of the PDU sessions and their 5QI (or QCI) profile, information about the DRB and the 5QI profile, the mapping of QoS flows to DRBs, the primary serving cell for the UE and the secondary cells (in case of Carrier Aggregation), the secondary node for the UE (in case of EN-DC or MR-DC), etc.
3. Configuration information
   This includes information pertaining to PDCP and RLC configuration for the UEs, PDCP duplication, cell selection/reselection priority for the UEs with respect to NR-specific Absolute Radio Frequency Channel Number (ARFCN) and EUTRA-specific EARFCN bands, MAC layer logical channel configuration, DRB split ratio, scheduler configuration information such as scheduling request periodicity, buffer status reporting periodicity, semi persistent scheduling periodicity, discontinuous reception (DRX) cycle periodicity, number of Hybrid Automatic Repeat Request (HARQ) processes, and Channel Quality Information (CQI) configuration information.
4. Network interface or RRC messages
   This includes reporting a copy of network interface messages between E2 nodes and RRC messages between the UE and the E2 node to the near-RT RIC.

The near-RT RIC can also exercise UE-level control actions back to the E2 nodes using E2SM-RC for functionalities that include (i) radio bearer control – such as controlling the QoS profile of the DRB, the mapping of QoS flows to the DRB, configuring the logical channel for the DRB, admission control for the DRB and the PDCP/RLC configuration, controlling the DRB termination, split ratio, and PDCP duplication; (ii) radio resource allocation control – such as DRX parameter configuration, scheduling request periodicity configuration, semi persistent scheduling periodicity control, grant configuration, etc.; (iii) connected mode mobility control – such as choice of the optimal target cell for UE handover, conditional handover for UEs, and DAPS to control the handover of URLLC UEs; (iv) radio access control – such as admission control for the UE in terms of PDU sessions, DRB configuration, RACH back off control, access bearing control, RRC connection release control, and RRC connection reject control; (v) dual connectivity control – in terms of choice of secondary node, PSCell, etc.; (vi) carrier aggregation control – in terms of choice of secondary cells; (vii) idle mode mobility control – in terms of choice of cell reselection priority; and (viii) measurement reporting configuration control – in terms of controlling the measurement objects, reporting objects, etc.

Thus, E2SM-RC helps in the exchange of fine-grained information involving the UEs, QoS flows, DRBs, PDU sessions, slices, cells, etc., at near-real-time granularities with the near-RT RIC.

5.6.3 Other E2SMs

Other E2SMs, standardized (or being standardized) in O-RAN WG3, include E2SM-Cell Configuration and Control (E2SM-CCC) and E2SM-Network Interface (E2SM-NI). E2SM-CCC enables reporting of cell-level and network element-level configuration information to the near-RT RIC and facilitates the near-RT RIC to control the configuration parameters associated with the network elements at near-real-time periodicity. On the other hand, E2SM-NI enables the tracing of UE-associated network interface messages from the E2 nodes to the near-RT RIC, thereby enabling the near-RT RIC to gain access to fine-grained information about UE state information exchanged as part of network interface procedures between the E2 nodes.

The E2SMs thus facilitate reporting the relevant UE and RAN data as state information to AI/ML engine in the near-RT RIC and also facilitate optimization of control parameters by the near-RT RIC xApps back to the E2 nodes.

5.7 ML Algorithms for Near-RT RIC

As discussed in Section 5.1, the evolution of RATs and newer use cases demand the necessity to use AI/ML algorithms that explore the complex and intricate interdependencies between the parameters, state, context, and performance information across the layers of the RAN protocol stack for optimizing the RRM decisions of the control variables toward meeting the target KPI objective. The parameters, state, context, and performance information generated from the E2 nodes as indications at UE-level/cell-level/node-level are considered input features to the ML engine, which outputs the RRM decisions for the control parameters sent as control actions from the near-RT RIC to the E2 nodes. The information sent via the indication messages from the E2 node to the near-RT RIC and the control actions sent from the near-RT RIC back to the E2 nodes are based on E2SMs, as discussed in Section 5.3.

It is to be noted that the near-RT RIC cannot do offline ML model training but can perform online ML model training and RL. The non-RT RIC has the required computational and storage capacity to perform offline ML model training. So, the near-RT RIC can request the non-RT RIC to build an ML model with offline training, which can later be downloaded in the near-RT RIC, where the model can be updated and deployed in the xApp that acts as an inference host. The general steps for building ML algorithms to facilitate intelligence in the RIC are as follows:

AI/ML model development using E2SM data: The E2 interface is used to send indications, containing relevant data, from the E2 node to the near-RT RIC using E2SMs. The xApp in the near-RT RIC shall access the data and may decide to request for AI/ML training services to the non-RT RIC for generating an offline-trained ML model. The corresponding rApp in the non-RT RIC shall receive this request. Once the non-RT RIC receives the request for AI/ML training service from the near-RT RIC, the rApp requests for data to be streamed from the near-RT RIC. The xApp in the near-RT RIC can use the O1 interface to send the E2SM data to the SMO at non-RT granularities. The SMO/non-RT RIC stores the E2SM data and the rApp in the non-RT RIC asks the SMO/non-RT RIC framework to perform offline AI/ML training towards building an offline ML model and provides the required hyperparameters, in the process, for the model training. The ML model is stored in a repository in the SMO/non-RT RIC framework and the model is uploaded to the catalog. Once available, the ML model is downloaded in the near-RT RIC over the O1 interface, and the details are responded back to the near-RT RIC over A1. The model is then deployed in the xApp that acts as the inference host. The ML model downloaded in the near-RT RIC can further be subject to updates via online training based on the data sent to the near-RT RIC from the E2 node. The updated ML model can be pushed to the SMO/non-RT RIC, and the updated model is stored in the repository with the details updated in the catalog. The updated ML model is then deployed in the xApp as the inference host.

AI/ML-driven A1 policy: The non-RT RIC rApps use the PM/KPI data received from the O-CU-CP, O-CU-UP, and O-DU over the O1 interface for developing AI/ML-driven A1 policy and enrichment information, which are sent to the near-RT RIC and consumed by the corresponding xApps. This A1 policy information is used by the near-RT RIC xApps to set the RRM objectives and targets for RRM.

5.7.1 Reinforcement Learning Models

Figure 5.3 illustrates how RL models can be built in the near-RT RIC. There are various categories of RL algorithms:

Model-free vs. model-based: Model-free RL algorithms do not model the state transition probability in the environment due to actions, but estimate the reward from state-action samples toward taking subsequent actions. Whereas model-based algorithms model the state transition probability to learn the inner workings of the environment toward predicting the optimal control actions, accordingly.

Off-policy vs. on-policy: In off-policy RL algorithms, the target policy (the policy that the RL agent is trying to learn to determine and subsequently improve its reward value function) is different from the behavior policy (the policy used by the RL agent to generate action toward interacting with the environment). Off-policy RL agent makes use of a replay buffer which consists of data samples from the environment pertaining to all prior policies toward generating a newer/updated policy. On-policy RL algorithms use the same policy for both target and behavior.

Offline vs. online: In offline RL, a fixed training dataset of logged experiences is collected based on any behavior policy, which could also be potentially unknown. The RL agent is trained without any interactions with the environment but based on this fixed offline training dataset of logged experiences. The policy is deployed online only after it is fully trained. Whereas in online RL, the agent interacts with the environment online and a policy is updated to a newer policy based on the streaming data from the environment collected by the policy itself.

Value-based RL vs. policy-based RL: In value-based RL, the values of the action candidates based on the state vector are computed by the RL agent, and the action with the best value is determined, whereas in policy-based RL, the RL agent learns the stochastic policy that maps the state vector to the action.

An RL agent can be present in the near-RT RIC framework or in the xApp. As seen in Figure 5.3, the INDICATIONS from the E2 nodes using the relevant E2SMs can constitute the state vector to the RL agent, and the state vector could be UE-specific. The E2 nodes and the underlying RAN

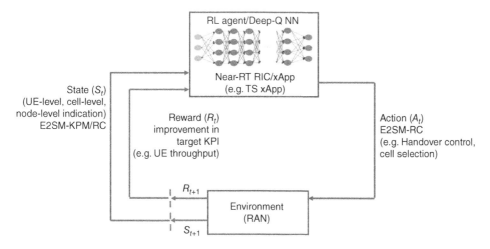

Figure 5.3 Illustration of a reinforcement learning engine in the near-RT RIC. The engine receives state information and computes rewards based on E2SMs that stream indication messages and generates control action to optimize RRM decisions based on E2SM.

constitute the environment. The non-RT RIC sets the KPI target and the objectives and are sent to the near-RT RIC via the A1 interface, and the RL agent computes the reward as an improvement in the target KPI, further computed from the INDICATION messages sent from the E2 node using E2SM. The E2SM CONTROL actions and the parameters constitute the actions taken by the RL agent in terms of optimizing the variables and decisions. Taking the example of an RL algorithm such as Deep-Q Network (DQN), the non-RT RIC can be leveraged for training the offline RL model for learning the reward as a function of the RAN data obtained from E2SM INDICATION messages and the optimization variables controlled by E2SM CONTROL actions. Once the offline ML model is downloaded in the near-RT RIC, the xApp can exploit the learnings of the ML model toward making inference on the control variables using E2SM, based on the incoming stream of INDICATION data from the E2 nodes. Moreover, the downloaded ML model can also be subject to policy updates via further explorations by interacting with the environment. The RL agent can generate a random control action that gets reflected in the environment toward generating INDICATION messages, which can be used to update the RL target policy in the RL agent. The updated model can then be subsequently deployed in the xApp, which makes further inferences based on the updated model. The updated model is also uploaded to the non-RT RIC. And the updates to the ML can continue with further exploration, until convergence that minimizes the loss function (standard Bellman error, in the case of Q learning).

It is to be noted that RL can be applied to systems that are modeled as Markov decision processes (MDPs), where the probability of transition of the current state vector to the new state vector is dependent on the current state vector and the action taken by the RL agent toward the transition to the new state vector. As an example, for the traffic steering O-RAN use case, the traffic steering xApp can make use of an RL agent that receives the UE-specific E2SM-KPM and E2SM-RC indication reports (containing UE context/state information and PMs, serving cells and serving E2 node context and PM information, UE's L3 RRC information for neighbor cells and neighbor cell context information, etc.) as the state vector from the E2 nodes (environment) to the near-RT RIC, and the RL agent in the xApp can generate a UE-specific handover control action that optimizes the decision of the target cell for the UE toward optimizing the mobility/handover decisions that maximize a given KPI target for the UE (such as throughput/latency). RL models are usually a good choice for closed-loop control systems such as the near-RT RIC for optimizing the RRM decisions in the E2 nodes.

5.8 Near-RT RIC Platform Functions for AI/ML Training

The near-RT RIC architecture, as standardized in O-RAN WG3 and as shown in Figure 5.4, includes the following platform functions for AI/ML training:

- **AI/ML support:** This platform function provisions the following services:
  Data pipeline: The AI/ML data pipeline service for the AI/ML support function in near-RT RIC offers data ingestion and preparation for xApps. The input to the AI/ML data pipeline may include E2 node data collected over E2 interface, enrichment information over A1 interface, information from xApps, and data retrieved from the near-RT RIC database through the messaging infrastructure. The output of the AI/ML data pipeline may be provided to the AI/ML training capability in near-RT RIC.
  Training: The AI/ML training service for the AI/ML support function in near-RT RIC offers training of xApps within near-RT RIC. The AI/ML training provides generic and use

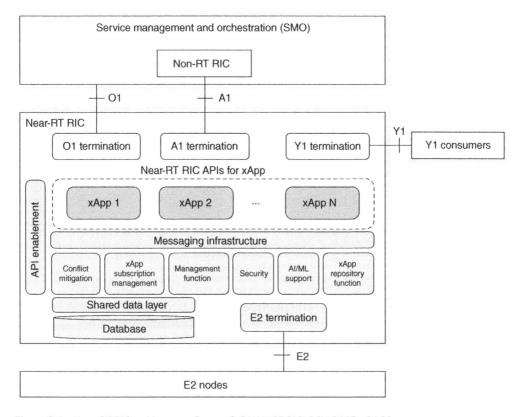

Figure 5.4 Near-RT RIC architecture, *Source:* O-RAN.WG3.RICARCH-R003-v04.00

case-independent capabilities to AI/ML-based applications that may be useful to multiple O-RAN use cases, such as traffic steering, QoS optimization, QoE enhancement, slicing, MU-MIMO, and RAN sharing.

- **API messaging infrastructure for AI/ML services:** O-RAN WG3 standardizes APIs and interfaces that enable the xApps to communicate with the AI/ML support functions in the near-RT RIC platform toward training AI/ML models and subsequent deployment of updated models in the xApps. These APIs are standardized so as to enable multi-vendor interoperability between third party xApps and the near-RT RIC platform that offers AI/ML services.

5.9 RIC Use Cases

There are many use cases that can be enabled by leveraging RIC and ML (see Figure 5.5). RIC presents a vehicle to achieve Total Cost of Ownership (TCO) reduction goals without vendor lock-in. As an open platform, operators can benefit from deploying xApps and rApps that use a plethora of datasets (derived internally from operating the RAN network but also externally that may be useful for the RIC to prioritize specific features), for example, packet core information, weather or traffic information, etc. An important point to keep in mind is the logical decoupling of the SMO framework into non-RT RIC, RAN O&M, and generic SMO functions. These functions can share data among each other and provide flexibility for MNOs while adhering to O-RAN principles.

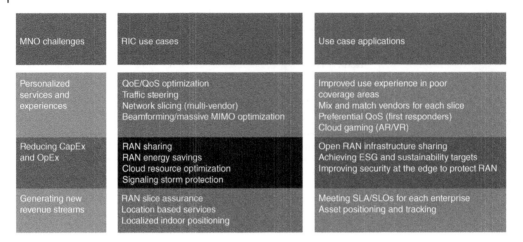

Figure 5.5 RIC use cases and applications

Typical use cases for TCO reduction include:

– **Automating new RF carrier and site launches**: As MNOs roll out 5G, an automated process to add new carriers and site launches is critical to OPEX savings. With the help of AI/ML, the full life cycle can be automated with the assistance of MDT (minimization of drive test – a standardized mechanism in 3GPP since Release 10).

A high-level process of a new site/carrier launch is shown below.

5.10 Conclusion

In this chapter, we broadly review the ML life cycle involving the RIC components, especially focusing on the near-RT RIC and non-RT RIC. We further reviewed E2SMs which serve as the basis for AI/ML realization in the near-RT RIC in terms of providing data to the RIC and facilitating control action generation from the RIC. We further discussed how AI/ML algorithms and RL models can be built in the RIC, and how ML life cycle management can be exercised in the RIC based on O-RAN specifications. We also provided details on the relevant AI/ML support standardized in the near-RT RIC and non-RT RIC framework by O-RAN WG3 and WG2 technical specifications. Finally, we briefly compared the legacy or present mode of operation for SON and automation and the future direction being led by the open RAN movement.

Bibliography

3GPP (2017). *3rd Generation Partnership Project; Technical Specification Group Radio Access Network; Study on New Radio Access Technology: Radio Access Architecture and Interfaces (Release 14)*.
3GPP (2021). *Summary of RAN Rel-18 Workshop*. O-RAN.WG2.AIML-v01.03 (2021). *O-RAN AI/ML workflow description and requirements*.
O-RAN.WG1.O-RAN-Architecture-Description-v06.00 (2022). *O-RAN Architecture Description*.
O-RAN.WG2.A1TS-v02.00 (2022). *O-RAN Working Group 2 A1 Interface: Test Specification*.
O-RAN.WG3.E2TS.v01.00 (2022). *O-RAN Working Group 3 E2 Interface Test Specification*.

6

The Fronthaul Interface

Aditya Chopra

Amazon Kuiper, Austin, TX, USA

6.1   The Lower-Layer Split RAN

The radio access network (RAN) can be classified into two fundamentally unique subsystems: the analog component (typically called the radio) that emits or senses electromagnetic signals in the radio spectrum, and the digital component (typically called the baseband) that translates these signals into information packets carrying user data and connection control information.

The difference in the radio and baseband components of the radio access network (RAN) also translates into fundamentally different deployment requirements for each of these subsystems. For example, in order to maximize the efficiency of electromagnetic emission and sensing, the radio must be deployed at a height with obstruction-free views of its service region. Hence, cellular radios are frequently deployed on towers spread out across cities and rural areas. In contrast, the baseband processing subsystems are primarily performing digital compute operations. Ideally these should be deployed at centralized locations with economical access to fiber-optic connectivity, space, power, and cooling.

The first few generations of cellular deployments used coaxial cables to transmit analog signals between the radio and the baseband units. Cabled analog transmission suffered from significant loss of signal strength as well as signal quality, especially at the increasingly high carrier frequency of deployments. This meant that the baseband and radio components were forced to be colocated, resulting in baseband systems being deployed at the base of cellular towers.

The analog interface between the radio and the baseband was replaced with digital signaling in 4G-LTE cellular deployments. This digital link between the baseband and the radio subsystems is commonly referred to as the "fronthaul". The baseband and radio would communicate via proprietary messaging protocols, carried via a digital transport interface known as the common public radio interface (CPRI). The CPRI transport interface was not a switched interface; in other words, a network operator could not put network routers between the baseband and the radio. Thus, while CPRI alleviated the problem of signal strength loss between the radio and baseband, it still did not relax the colocation requirement (Figure 6.1).

An upgrade to CPRI, the enhanced CPRI, or eCPRI interface was defined as a packet-switched interface in (CPRI 2019). The eCPRI interface allows packet switches to be present between the radio and baseband subsystems, finally allowing geographic separation in their deployments as

Open RAN: The Definitive Guide, First Edition. Edited by Ian C. Wong, Aditya Chopra, Sridhar Rajagopal, and Rittwik Jana.
© 2024 The Institute of Electrical and Electronics Engineers, Inc. Published 2024 by John Wiley & Sons, Inc.

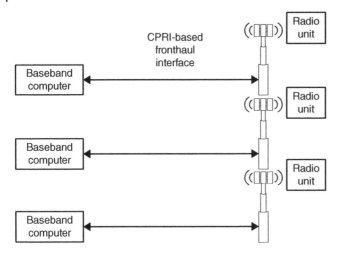

Figure 6.1 CPRI-based fronthaul architecture.

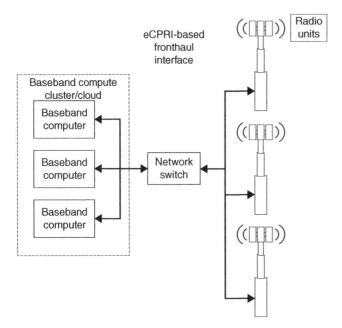

Figure 6.2 eCPRI-based fronthaul architecture.

shown in Figure 6.2. While both the CPRI and the eCPRI interfaces are open specifications, they primarily focus on describing packet headers and interfaces, and do not specify the messages between the radio and baseband. This allowed radio vendors to use proprietary messaging within CPRI to achieve desirable radio and baseband separation.

6.1.1 Lower Layer Fronthaul Split Options

With key deployment benefits offered by a packetized digital fronthaul, it becomes imperative for any Open RAN standards to create an open fronthaul specification that allows interoperability between different radio and baseband vendors. The O-RAN Alliance was not the first such standards body to

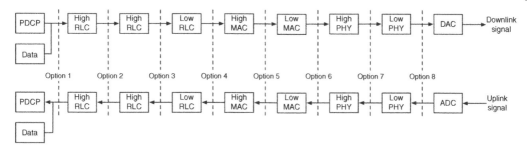

Figure 6.3 RAN split options proposed by 3GPP.

attempt to standardize the split RAN architecture, but was the first to standardize the fronthaul interface and its corresponding messages. In 2015, the Third Generation Partnership Project (3GPP) undertook a study item to standardize split architecture on cellular base stations. While these standardization efforts concluded at the end of the study phase, the resulting technical report (3GPP 2017) did provide a nomenclature for different points within the data processing chain where a split may be introduced. Figure 6.3 shows the named split options within the baseband processing chain of a cellular RAN. In this figure, the term "PHY" stands for Physical Layer, and indicates the digital signal processing components of the RAN. "MAC" is an abbreviation of Medium Access Control and contains aspects related to error correction coding, retransmission, packet acknowledgments, and multiuser access. RLC indicates "Radio Link Control" which optimizes the concatenation and segmentation of user packets, error detection, and recovery. The High and Low terms typically indicate command-, control-, and overall optimization-related features and high-speed repetitive data processing.

Since the fronthaul separates the radio and the baseband, the candidate split options for a lower layer split were limited to Options 6, 7, and 8, which occur around the functional components that operate at the physical layer. While choosing the split, the following key tradeoff must be taken into consideration:

a) A split fronthaul interface at a "lower level" (toward the right in Figure 6.3) keeps the radio unit (RU) as simple as possible in terms of lower size, weight, and power draw, thereby dramatically reducing the cost for tower top deployments.
b) A split fronthaul interface at a "higher level" (toward the left in Figure 6.3) reduces the fronthaul throughput requirements, thereby simplifying the cost and complexity of the transport network needed to support the fronthaul.

The remainder of this chapter investigates the three lower layer splits and the well-known specifications that use these split planes. In order to maintain consistency in terminology, the baseband processing system at one end of the fronthaul split is labeled the Distributed Unit or DU, and the radio system on the other end of the fronthaul split is labeled the Radio Unit or RU. The analog front-end components always exist inside the RU, while the digital signal processing is distributed among the DU and RU depending on the split option. In the section discussing the O-RAN Option 7-2x split, the DU is labeled as the O-RAN DU or O-DU, and the RU is labeled the O-RAN RU or O-RU.

6.2 Option 8 Split – CPRI and eCPRI

The Option 8 split separates the baseband and the RU at their analog-to-digital interface. In this split, a steady stream of time-domain digital signal samples is transferred between the RU and the baseband. On the downlink, the baseband sends a digitally encoded stream of signal sample values

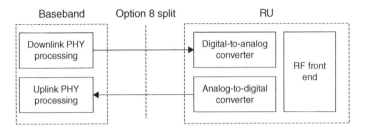

Figure 6.4 Option 8 fronthaul split between baseband and radio unit.

to the RU, which converts them into analog voltage for subsequent RF transmission, whereas on the uplink, the RU receives RF signals, digitizes them, and transfers the resultant digital data to the baseband (Figure 6.4).

Historically, the CPRI specification was designed to carry such data between the RU and the baseband and widely used as the fronthaul interface for 4G cellular network deployments. It provided a clean separation of analog and digital signal processing components in the network, and while the lack of signal processing at the RU meant that raw wireless signals were being transferred across the split, the fronthaul throughput requirements were manageable for the bandwidth used by 4G carriers.

In 5G cellular networks, especially millimeter wave deployments with extremely large bandwidths, and midband massive multiple-input multiple-output (MIMO) deployments with large numbers of antennas, the amount of data transferred across the fronthaul can be onerous. A simplistic serial data transfer such as CPRI would not scale well in such deployment scenarios, leading to the creation of enhanced CPRI (eCPRI). eCPRI uses an ethernet-based transport layer that allows the use of intelligent packet routing. The eCPRI specification also allows the use of data compression and signal processing at the RU to manage fronthaul throughput. While eCPRI only publishes an Option 8-based signal data-encoding specification, it allows network equipment vendors to send proprietary messages between the baseband and the RU using its transport layer specification. In fact, the O-RAN Alliance Open Fronthaul uses the eCPRI specification as its transport layer to exchange Split Option 7-based messages between its baseband and RU.

The Radio over Ethernet (RoE) specification published by the Institute of Electronic and Electrical Engineers (IEEE) is also similar to eCPRI in that it is primarily an Option 8-based fronthaul specification that allows for proprietary data compression and control messages. IEEE RoE can also encapsulate and carry both CPRI and eCPRI data packets.

6.3 Option 6 Split – FAPI and nFAPI

While the Option 8 split occurs between the RF and the PHY, the Option 6 split occurs between the PHY and the medium access control (MAC). In an Option 6 split RAN, the entirety of PHY processing is placed inside the RU. The data communicated between the PHY and the MAC is purely user-generated information bits. Therefore, the Option 6 split can be considered as the most efficient lower layer split in terms of fronthaul throughput requirements since it only communicates the bits that are sent to and from the user. However, there is a catch, which is that this efficiency comes at the cost of performing all of the PHY compute within the RU. The same type of deployments that are disadvantageous to Option 8 splits, namely high bandwidth mmWave and high antenna count massive MIMO, also require large amounts of PHY computation and hence are disadvantageous for Option 6 splits as well. This can significantly drive up the cost, size, and complexity of the RU. Furthermore, as standardization progresses through multiple releases, there may be new features that require modification to PHY-processing algorithms. The development and deployment costs of new features can be significantly higher on RU platforms compared to general-purpose baseband processing platforms. While there are issues with using an Option 6 split in macro deployments described earlier, some of these issues are suppressed in small-cell indoor deployments. Given the limited reach and low number of users per cell in such deployments, a low-complexity PHY can be deployed by the RU.

An open implementation of this split was originally specified by the small cell forum (SCF) as an intra-baseband hardware application programming interface (API) for controlling special-purpose PHY signal processing hardware by a general-purpose computer running the MAC layer. An example of this would be a digital signal processor or field programming gate array (FPGA) on a PCI Express card executing the PHY layer in a server with a 64-bit general-purpose CPU executing the MAC layer. This API was called femto application platform interface (FAPI), and over time, SCF extended the FAPI specification to geographically split processing of the PHY and the MAC. The new specification essentially carried the original FAPI commands over a networked interface and was hence known as nFAPI for networked FAPI. Figure 6.5 shows the FAPI split architecture and Figure 6.6 shows the nFAPI split architecture. This table summarizes the differences between FAPI and nFAPI.

FAPI	nFAPI
Internal interface within a small cell product	External (network) interface within a small cell network composed of centralized and distributed units
1 MAC – 1 PHY (1–1 mapping)	1 MAC – N PHYs (1-many mapping)
Physical transport typically has high bandwidth, low latency, and low jitter (i.e. ideal transport)	Network transport can be packet-based (e.g. ethernet) and with nonideal bandwidths, latencies, and jitter.
Synchronization between MAC and PHY easier	Synchronization between MAC (CU) and PHY (DU) more critical

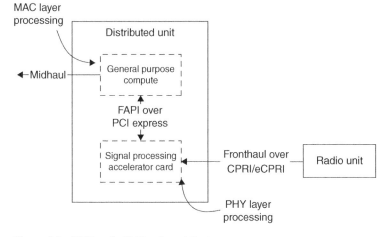

Figure 6.5 FAPI and nFAPI split architectures.

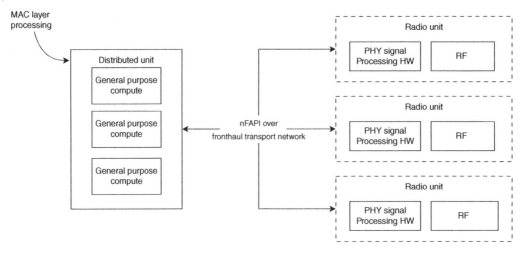

Figure 6.5 (Continued)

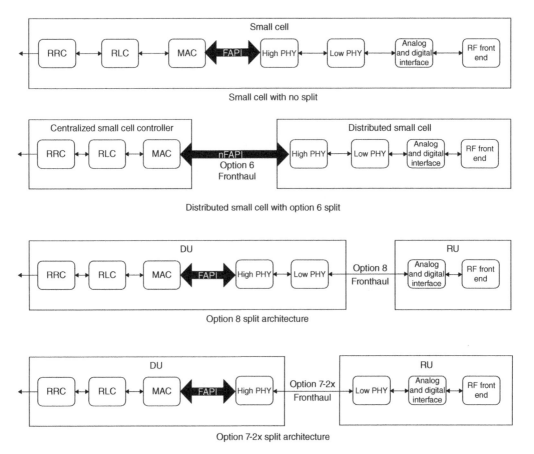

Figure 6.6 Using FAPI and nFAPI in various split RAN architectures.

6.3.1 Subinterfaces

The FAPI and nFAPI specifications contain a set of open interfaces, namely P7 for the main data path and P5 for the control path between the MAC and the PHY layers. The next important Open RAN interface is the "RF/Digital front end" interface, which has been specially introduced for enhanced 5G capabilities, namely MIMO and advanced beam-forming capabilities. This interface is termed P19 and interfaces MAC to the RF/Digital front end. This was not needed for the earlier generations because the RF front ends were simple and did not need/allow sophisticated control.

More recently, the P4 interface, which is also referred to as the "Network Monitor Mode" interface, has been introduced. This allows the small cell to listen to the radio network environment and report its observations to the upper layers of the RAN protocol stack for sophisticated self-organizing network (SON)-like functions.

6.3.2 Architecture Agnostic Deployment

Even though these were developed for Split-6 solutions, it is worth pointing out that they, in particular, FAPI, are critical components even for the CPRI or eCPRI-based Option 8, or O-RAN alliance-defined Split-7-2x solutions. FAPI is essentially an Open Interface between the PHY and MAC layers of the RAN protocol stack. As such, this interface is crucial as an internal interface within the integrated gNB or 3GPP-defined gNB-DU or O-RAN alliance-defined O-DU. Figure 6.6 shows the different locations within the RAN where FAPI and nFAPI can be deployed for different types of split architectures.

6.4 Option 7 Split – O-RAN Alliance Open Fronthaul

Option 7 from Figure 6.4 is a unique split option that splits baseband processing between the DU and the RU. It is also referred to as split PHY as the physical layer processing functions are split within the DU and the RU. In Option 8, all baseband physical layer processing happens in the DU, while in split 6, all baseband physical layer processing happens at the RU. Since this option splits within the baseband processing, there are multiple subsplits that can occur depending on where the baseband processing is split. Each of these has advantages and disadvantages.

The xRAN alliance in 2017 released the first version of an open fronthaul specification based on a 7-2 split. After xRAN was merged into O-RAN in 2018, this specification became the O-RAN open fronthaul specification, with the fronthaul split plane designated as "7-2x". Figure 6.7 shows the distribution of PHY functions between the O-RU and the O-DU across this split plane. The key benefit of this split design over the 7-2 split is that the O-RU on the uplink transmits "streams" across the fronthaul. The streams are beamformed reductions of the antenna chains received by the O-RU. Typically, in Option 8 split, the antenna chains are transmitted across the fronthaul, while in Option 7, user layers are transmitted across the fronthaul. "Streams" lie between the two and can scale to either end. This flexibility is the key power of the O-RAN split option 7-2x and allows the RAN to scale up or down the amount of information received from the O-RU depending on the user and channel conditions (Figure 6.7).

The O-RAN Alliance established nomenclature for the various nodes and interfaces within their openly specified split RAN architecture. The node containing RF emitting and sensing antennas and analog signal processing elements is called the Open RAN RU or O-RU. The O-RU also contains some digital signal processing capabilities to convert back and forth between analog

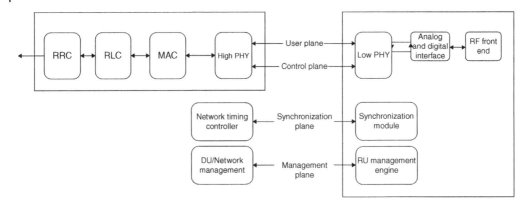

Figure 6.7 Option 7-2 split architecture.

RF signals and digitized information packets. The computational node that processes the baseband packets is known as the Open RAN-distributed unit or O-DU.

The fronthaul interface connecting the O-DU to the O-RU is divided into multiple subinterfaces, also known as "planes". The "control", "user", and "synchronization" planes are specified in the O-RAN FH CUS specification (O-RAN 2022), while the "management" plane is specified in the O-RAN FH M specification (O-RAN 2022). The remainder of this section describes each of the planes.

6.4.1 Control (C) and User (U) Plane

The control and user plane are responsible for downlink and uplink user data transfer between the O-DU and the O-RU. The user plane carries cellular user data in the form of frequency domain I and Q samples. It is named so because these messages essentially encoded communication data packets to and from cellular subscribers. The user plane is bidirectional and carries downlink radio signals from the DU to the RU, and uplink radio signals from the O-RU to the O-DU.

The control plane is responsible for messages related to O-RU preparation for downlink data packets, requesting uplink data of interest, analog, and digital beamforming as well as beam weight exchange, and setting other real-time parameters on the O-RU such as compression. This subinterface carries messages that control O-RU behavior. It is a unidirectional interface that exchanges messages from the O-DU to the O-RU. Control plane messages are used to indicate to the RU the type of user plane packets to expect in the downlink direction in the near future, as well as the time and frequency resource location of future uplink signals. Control plane messages also control real-time O-RU parameters such as digital and analog beamforming.

O-RAN Open Fronthaul provides the ability to use either eCPRI or IEEE RoE transport headers to encapsulate control and user plane packets. This allows commercial fronthaul gateway equipment that supports eCPRI and RoE to interpret the packet headers and appropriately prioritize and route these packets. Fronthaul packets are Ethernet addressable and can be used with VLAN tagging for use in virtual RAN deployments.

6.4.2 Management (M) Plane

The management plane carries messages that control O-RU parameters such as center frequency, bandwidth, and analog power levels, as well as monitor the operational status of the O-RU. The management plane is also differentiated by its speed and latency compared to the control and user planes. The majority of management plane messages do not have the strict latency or response time requirements of the control and user plane and thus are typically handled by different processing components in the RU. For example, the control and user plane messages typically must be handled by special hardware such as onboard DSPs and/or FPGAs, while management plane messages are typically processed by onboard CPUs.

In the O-RAN architecture, management of the O-RU is handled by two entities – the DU and the SMO. There is another management interface between the SMO and the RU, denoted by "O1". Work is ongoing to determine the distribution of scope between these two interfaces.

The management (M) plane in O-RAN Open Fronthaul specification describes the protocols and messages exchanged over the fronthaul interface linking the O-RU with other management plane entities, which may include the O-DU, the O-RAN-defined service management and orchestration (SMO) as well as other generic network management systems (NMSs). The M-plane facilitates the initialization, configuration, and management of the O-RU to support the transmission of user and control data over the functional split.

A NETCONF/YANG-based M-Plane is used for supporting the management features including "start-up" installation, software management, configuration management, performance management, fault management, and file management toward the O-RU. The M-Plane supports the following two architectural modes:

1) **Hierarchical model:** This M-plane model is shown on the left side (Figure 6.8). Here an O-RU is managed entirely by a single O-DU using a NETCONF-based M-plane interface.
2) **Hybrid model:** This M-plane model is shown on the right side of Figure 6.8. Here the hybrid architecture allows multiple interfaces between the O-RU and the network. One interface could be between the NMS and the O-RU, while in parallel, there may be a logical interface between O-DU and the O-RU. The interfaces can also separate management responsibilities, for example, the network management interface may control functions like O-RU software management, performance management, configuration management, and fault management, while the O-DU interface controls the real-time parameters of the fronthaul link.

In the hybrid model, the O-RU has end-to-end IP layer connectivity with network management also known as SMO.

NETCONF/YANG was chosen as a standardized network element management protocol and data modeling language in order to simplify integration between O-DU and O-RU as well as

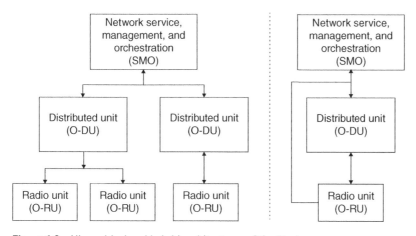

Figure 6.8 Hierarchical and hybrid architectures of the M-plane.

operator network integration. The framework supports the integration of products with differing capabilities enabled by well-defined data models that are published within O-RAN M-plane specifications. NETCONF also natively supports a hybrid architecture which enables multiple clients to subscribe and receive information originating at the NETCONF server in the O-RU.

Based on the transport topology, various modes of network connectivity are possible between O-RU and O-DU, and SMO. The basic requirement for M-Plane is to have end-to-end IP connectivity via IPv4 or IPv6 between the O-RU and the elements managing it such as the O-DU or SMO.

6.4.3 Synchronization (S) Plane

The synchronization plane carries messages to establish timing synchronization between the O-DU and multiple O-RUs. The details of the S-plane and overall synchronization in an Open RAN architecture can be found in Chapter 7 – Transport and Synchronization.

Figure 6.9 shows the nodes and the interface planes that form a split fronthaul-based RAN solution. The directionality of these planes is also highlighted.

6.4.4 Key Features

6.4.4.1 Fronthaul Compression

Earlier in this chapter, we saw that the choice of the fronthaul split option is based upon the fundamental tradeoff between RAN communication performance and fronthaul throughput constraints. However, the modulation of this tradeoff point does not have to stop once the split option has been selected. Even within a chosen split option, fronthaul data compression is another tool available to RAN fronthaul designers wanting to optimize the service quality to end-users for a given network deployment.

The bulk of the data transmitted across the fronthaul is carried by the user and control planes, and its information content (which is directly related to its compressibility) is highly dependent on the type of deployment that the fronthaul is serving. Consequently, the nature of RAN deployment has a large impact on the type of compression methods that are optimal for reducing fronthaul throughput requirements while maintaining an acceptable level of throughput.

It was this notion of "one size does not fit all" in compression approaches that led to the O-RAN fronthaul being designed to support multiple compression methods, with compression parameters tunable in real-time. In order to make such a design feasible, the O-RAN specification defines compression headers within fronthaul messages. These compression headers contain the relevant parameters that provide the necessary information for the message receiver to compress or decompress IQ data. For downlink user plane messages, the O-DU determines the best way to compress

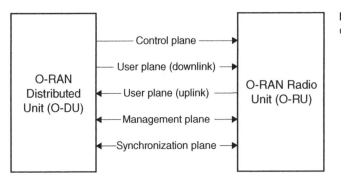

Figure 6.9 Option 7-2 split data flows.

IQ data and sends the compressed data along with the corresponding compression headers. The O-DU also decides the compression method for uplink data as well. Hence, it sends the compression header information inside the control plane message requesting an uplink resource block. The O-RU must compress uplink signals using the compression method and parameters provided by the O-DU. The O-RAN fronthaul allows the O-DU to select different compression methods and their parameter values for different resource blocks within a cell.

The key compression methods available in the O-RAN fronthaul specification are:

- **Fixed Point compression**
  Fixed point compression is the computationally simplest method of reducing throughput of fronthaul packets. This "compression" is performed by selecting a bitwidth for I and Q data samples and keeping only the most significant bits of each sample. This method of compression is clearly lossy as it discards the least significant bits of I and Q samples, thereby reducing the dynamic range of the compressed signal. In uplink communications, signals received at the basestation from cell edge users are likely to occupy the least significant bits of the digitized I and Q data samples. Discarding these bits for the purpose of fronthaul compression can negatively impact uplink communication performance of cell edge users.

- **Block Floating compression**
  Block Floating compression seeks to alleviate the core issue of reduced dynamic range seen in Fixed Point compression. Here the signal to be compressed is collected into blocks of 12 I and 12 Q samples, which also corresponds to a physical resource block in cellular communication standards. Each of the 12 I and 12 Q samples are shifted by a single shared exponent, and the resulting mantissa is discretized to a small number of bits. The shared exponent and the mantissa are transmitted across the fronthaul and block floating compression can achieve significantly improved compression performance compared to Fixed Point. A key caveat to note is that, within a block, the dynamic range of Block Floating compression is identical to Fixed Point compression for the same number of mantissa bitwidth, but across blocks, the dynamic range of Block Floating compression method is much higher.

- **Block Scaling compression**
  Block Scaling compression method is targeted primarily toward downlink fronthaul communication. Instead of a shared exponent used by Block Floating compression, Block Scaling uses a shared multiplicative scalar. In downlink communications, Option 7-2x signals are quadrature amplitude modulated data which are typically placed on a constellation grid. Once the scaling coefficient is removed from such signals, the remainder can be represented by a bitwidth that is one higher than the modulation order of the downlink signal. Thus, Block Scaling compression method approaches the maximum possible compression efficiency for a constellation based signal.

- **Modulation compression**
  Modulation compression is an enhanced version of the Block Scaling method. In Modulation compression, an additional constellation shift parameter is also transferred across the fronthaul, along with a block scalar and the compressed I and Q signals. The constellation shift parameter allows a quadrature amplitude modulated signal to be compressed using a bitwidth equal to the modulation order of the signal. Modulation compression can encode quadrature amplitude modulated signals using the minimum possible number of bits needed to preserve their information, compared to block scaling compression which uses one extra bit.

Additional details about these and other compression methods, as well as example code implementation of these compression methods can be reviewed in (O-RAN 2022).

6.4.4.2 Delay Management

Low latency and latency sensitive applications such as augmented and virtual reality, factory automation, and vehicular communication are becoming increasingly popular on 5G deployments. Fronthaul latency is a key contributor to overall system latency, and may have significant impact on the performance of a split architecture RAN supporting latency sensitive applications. Intra-PHY splits in particular, such as Option 7-2x and Option 8 splits, need fronthaul networks to meet stringent latency and transport jitter requirements. The O-RAN Open Fronthaul specification (O-RAN 2022) dedicates an entire chapter to this notion of delay management where downlink and uplink communication is defined in terms of fronthaul packet ingress and egress timings. Based on these definitions, the O-DU and O-RU are ascribed "delay categories" and "delay subcategories" for the purposes of allowing a matching of O-DU and O-RU units that will operate properly together from the point of view of accommodating a customer's fronthaul network characteristics. Network characteristics comprises the "time-of-flight" of signals through (typically) a fiber-optic cable (so can be estimated from the fiber length) added to the signal traversal latency through any switches within the fronthaul network. Pre-defined latency is necessary when actual latency measurements are not provided; both the use of predefined latency value and use of a method for measuring actual network latency in the DL and UL are supported in the O-RAN Fronthaul specification.

The delay category and delay subcategory values also depend in part on the processing latency and buffer sizes within the O-DU and O-RU. It may be expected, especially for an O-RU that the processing latency may depend on the specific frequency band and subcarrier spacing that is used. Further, a multiband radio may experience different processing latencies for different bands. Therefore, it may be expected that an O-RU (and perhaps more rarely an O-DU) will have different delay category and delay subcategory ratings for different bands.

6.4.4.3 Beamforming

With the accelerating adoption of higher bands of spectrum such as the midband (>2 GHz) and millimeter wave (>20 GHz), beamforming has become a critically important toolset for RAN vendors to overcome high propagation losses in order to provide large throughputs to end customers. Beamforming uses large antenna arrays at the RAN transmitter and receiver where the signal to and from each array gets manipulated in order to create an antenna pattern that is pointed at a particular user. There are two main domains in this signal manipulation, or beamforming is executed, digital-domain and analog-domain.

In digital domain beamforming, a low-dimensional signal is expanded into a larger dimension through multiplication with a matrix. In orthogonal frequency-division multiplexing (OFDM)-based communication systems, this operation can be done in frequency domain (before the inverse fast Fourier transform (IFFT) stage of the transmitter) or in time domain (after the IFFT stage of the transmitter) or both. Frequency domain beamforming in OFDM allows different users to use the same time slot with different beams. In contrast, with time-domain beamforming, all the users and signals in a time slot use the same beam. Hybrid beamforming allows different users in the same time slot to use different beams (in the frequency domain) at the same time as all the users using a shared time-domain beam. An example is the case where the time-domain beam provides directivity in the elevation plane (so all users use the same elevation beam) while the frequency-domain beams provide directivity in the azimuth plane (so different users may use different azimuth beams). Digital beamforming is used in midband RAN deployments and is also referred to as massive MIMO.

In analog domain beamforming, the signal expansion or combination is conducted by analog gain and phase manipulation circuitry. These systems are wideband in nature and therefore only allow a single beam for a given time slot. Analog beamforming systems are usually directional in nature and are typically used in mmWave systems owing to mmWave arrays with large number of antenna elements in mmWave arrays not supporting fully digital beamforming for cost control.

The O-RAN Open Fronthaul supports both digital and analog beamforming via multiple methods of message exchange between the O-DU and the O-RU.

6.4.4.3.1 Analog Beamforming In O-RUs with analog beamforming, an index called "beamId" indicates the specific beam predefined in the O-RU to use. The beamId could indicate a frequency-domain beam or a time-domain beam or a combination of both ("hybrid" beam) and the O-DU needs to know to ensure the beamId is properly applied, e.g. the O-DU could not apply different time-domain beams to the different PRBs in the same OFDM symbol. The method the O-RU uses to generate the beam is otherwise not relevant; it could use the application of gain and phase controls on separate antenna elements, or use multiple energy-shaping antennas, or any other technology. The O-RU is expected to convey to the O-DU via the M-Plane on startup beam characteristics but the O-DU remains ignorant regarding how the beam is actually created by the O-RU.

6.4.4.3.2 Digital Beamforming and Massive MIMO Here the O-DU is meant to generate weights that create the beam, so the O-DU needs to know the specific antenna characteristics of the O-RU including how many antenna elements are present in the vertical and horizontal directions and the antenna element spacing, among other properties. Due to the large number of coefficients that may have to be transferred across the fronthaul, various beam coefficient compression methods have been discussed in (O-RAN 2022) to reduce the throughput requirements of beam coefficient transfer. O-RAN systems using digital beamforming require the transmission of many beam coefficient vectors for both downlink and uplink communication. The beam coefficient transmission can be a significant component of the overall fronthaul throughput, consequently, O-RAN specifications also provide compression methods for beam coefficients. While IQ data compression methods such as block floating are available to use for beam coefficient compression, targeted algorithms such as beam-space compression can provide even higher compression efficiencies. These compression methods are outlined in (O-RAN 2022).

6.4.4.3.3 Channel-Information-Based Beamforming In this method of beamforming, the O-DU provides the channel state estimates and scheduling information of specific scheduled users to the O-RU. Using these two informational elements the O-RU is able to calculate capacity maximizing beamforming weights for co-scheduled users. Compared to other beamforming methods, channel-information-based beamforming requires complex beamforming coefficient generation algorithms to be present at the O-RU.

6.4.4.4 Initial Access
Initial access in cellular networks comprises the protocols and procedures by which a user device detects the presence of the network and enters into a connected state within the network. During initial access, user devices have not achieved timing synchronization with the network, and initial access protocols and messages are designed with this in mind. However, this lack of time synchronization may pose a challenge for fronthaul splits to be able to perform optimally during initial acess.

For an Option 8-based fronthaul, initial access data is the same as any other type of data channel since it is all just a digitized representation of RF signals. Option 6-based fronthauls do not even carry initial access signals over the interface, instead relying on the RU to perform all uplink signal processing.

Initial access can be complicated in an Option 7-2-based split such as the O-RAN Open Fronthaul. The O-RAN split exchanges frequency domain signals over the fronthaul, which has the inherent implication that the RU is synchronized with the user device. This is not true during initial access. Hence, the frequency domain conversion of initial access channels is executed with special start times and durations exchanged via the control plane. The corresponding uplink received signals are sent to the O-DU via the user plane, with special PRACH message formats specified by O-RAN (2022).

6.4.4.5 License Assisted Access and Spectrum Sharing

Due to the paucity of spectrum in the lower bands, new access policies have allowed cellular networks to coexist and share spectrum in the unlicensed bands. For coexistence, the cellular networks must first sense the spectrum before executing a transmission within the unlicensed band. Thus, the fronthaul must also be aware of such procedures. The O-RAN open fronthaul specification (O-RAN 2022) provides messages in place for spectrum sensing in unlicensed environments, as well as policies for dropping downlink messages if the presence of another device was sensed before transmission.

6.5 Conclusions

The physical layer in the current 5G RAN is a high-complexity system processing large volumes of digitized wireless signals with deep design integration between analog and digital components. RAN system designers are faced with the difficult task of extracting energy and spectral efficiency from baseband designs by means of splitting and centralizing. Standard bodies with the collective experience and knowledge of global telecom vendors and operators are able to solve this problem by creating open fronthaul specifications that use intelligent and creative methods to create optimized lower-layer RAN architectures of the future.

Bibliography

3GPP TR 38.801 (2017). *3GPP Technical Report 38.801, Study on New Radio Access Technology: Radio Access Architecture and Interfaces, V14.0.0.*

O-RAN-WG4.CUS.0-v08.00 (2022). *O-RAN Fronthaul Working Group Control, User and Synchronization Plane Specification.*

O-RAN-WG4.MP.0-v09.00 (2022). *O-RAN Fronthaul Working Group Management Plane Specification.*

eCPRI Specification V2.0 (2019). Common *Public* Radio Interface:eCPRI Interface Specification

7

Open Transport

Reza Vaez-Ghaemi[1] and Luis Manuel Contreras Murillo[2]

[1] *VIAVI Solutions, Germantown, MD, USA*
[2] *Telefonica, Distrito Telefónica, MAD, Spain*

7.1 Introduction

Wireless transport networks have been essential in connecting a large number of cell sites. The previous generation of wireless services relied on Backhaul networks interconnecting these sites through central offices that also hosted elements of the core network. The adoption of smart phones drastically increased the demand for bandwidth and drove the introduction of technologies such as centralized radio access networks (CRANs). The latter pushed many active elements (baseband unit or BBU) of the radio networks into central locations and fueled the need for Fronthaul networks. 5G services are contributing to a new split architecture which centralizes a subset of active components (central unit or CU) in remote locations farther away and introduces a new network segment known as Midhaul. This chapter characterizes the requirements and solutions for Backhaul, Midhaul, and Fronthaul networks.

7.2 Requirements

Xhaul Transport and Synchronization networks represent the parts of the network that provide transport and synchronization functions for Fronthaul, Midhaul, and Backhaul networks. Figure 7.1 illustrates a simplified aspect of transport networks composed of a number of transport network elements (TNEs) inserted between RAN components such as O-RU, O-DU, and O-CU, and the core networks. Fronthaul networks can also include equipment or functions such as Fronthaul Gateway (FHG), Radio over Ethernet (RoE), Fronthaul Multiplexers (FHM), and cascaded radios. O-RAN Xhaul Transport Networks only consider Fronthaul Interfaces based on split option 7-2x as defined in O-RAN WG4 Open Fronthaul Interface Working Group 4.

Figure 7.1 contains two TNEs in a point-to-point topology in each network segment for the purpose of simplification. Real networks certainly contain multiple TNEs in a variety of topologies

Open RAN: The Definitive Guide, First Edition. Edited by Ian C. Wong, Aditya Chopra, Sridhar Rajagopal, and Rittwik Jana.

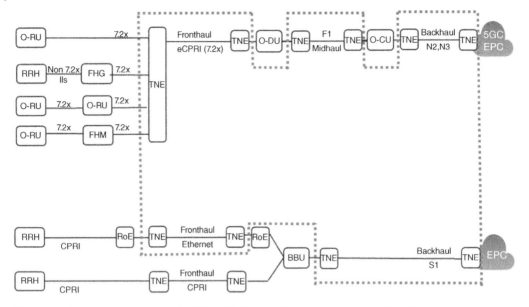

Figure 7.1 Xhaul transport networks [XPSAAS]. *Source:* With permission of O-RAN Alliance.

that are driven by the network requirements provided in this section. The Xhaul network requirements are broken into multiple categories:

– Fronthaul
– Midhaul and Backhaul
– Synchronization

The following sections contain some highlights of the requirements. For a complete set of requirements, please refer to (O-RAN.WG9.XTRP-REQ-v01.00 2020).

7.2.1 Fronthaul Requirements

Fronthaul requirements are driven by use cases with eMBB and URLLC representing the dominant cases. Frame loss ratio and one-way delay requirements are listed in Tables 7.1 and 7.2.

7.2.2 Midhaul Requirements

There are many similarities between Midhaul and Backhaul transport requirements. The following bullets highlight the main categories of requirements unique to Midhaul. There are some requirements that apply equally to Midhaul and Backhaul. The latter are covered in the next

Table 7.1 Frame loss ratio requirements.

CoS name	Example use	Maximum one-way frame loss ratio performance
High	User plane (fast)	10^{-7}
Medium	User plane (slow), C&M plane (fast)	10^{-7}
Low	C&M plane	10^{-6}

Table 7.2 One-way delay requirements [O-RAN.WG9.XTRP-REQ-v01.00 2020].

Latency class	Max. one-way frame delay performance	Use case
High25	25 μs	Ultra-low latency performance
High75[a]	75 μs	For full NR performance with fiber lengths in 10 km range
High100	100 μs	For standard NR performance with fiber lengths in 10 km range
High200	200 μs	For installations with fiber lengths in 30 km range
High500	500 μs	Large latency installations >30 km
Medium	User plane (slow) and C&M plane (fast)	1 ms
Low	C&M plane	100 ms

[a] New requirement category added based on deployment needs.

section. For details of the below categories, please refer to Xhaul Transport Requirement document, July 2020.

- 3GPP interfaces
- Network and transport protocols
- Logical connectivity
- Scalability
- Provisioning
- One-way delay

7.2.3 Backhaul Requirements

The following bullets capture the main categories of requirements that apply to Backhaul and Midhaul. For details of the these categories, please refer to [REQ].

- 3GPP interface
- Network and transport protocols
- Logical connectivity
- Interoperability
- Scalability
- Transport infrastructure
- Slicing capabilities
- Peak 5G downlink data rates and encapsulations
- Transport provisioning
- Transport technology
- One-way delay

7.2.4 Synchronization Requirements

3GPP has defined the following synchronization metrics for the air interface (Figure 7.2):

- **Time alignment error (TAE) relative:** is the time difference between any two antennas
- **Time alignment error (TAE) absolute:** is the time difference between an antenna port and the primary reference time clock (PRTC)

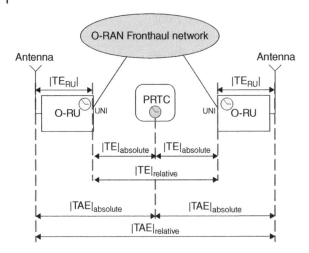

Figure 7.2 Definition of absolute and relative TAE and TE. *Source:* With permission of O-RAN Alliance.

Table 7.3 TAE limits for category A+ through C services.

Category	|TAE| absolute	|TAE| relative	Application
A+	N/A	65 ns	MIMO or TX diversity transmissions, at each carrier frequency
A	N/A	130 ns	E-UTRA intra-band contiguous carrier aggregation
B	N/A	260 ns	NR intra and inter-band contiguous carrier aggregation; E-UTRA intra-band noncontiguous carrier aggregation
C	1.5 µs	3 µs	NR intra- and inter-band noncontiguous carrier aggregation; TDD use cases

O-RAN uses the following metrics for the Xhaul network interface:

- **Time error (TE) absolute:** is the time difference between a user network interface (UNI) and the PRTC
- **Time error (TE) relative:** is the time difference between two UNIs

The relative and absolute TAE limits are dependent on the type of service. Table 7.3 provides limits for a number of 5G services.

7.3 WDM Solutions

This section is intended to describe best practices for O-RAN Fronthaul transport based on WDM technology. It is recognized that other solutions, not based on WDM technology, could be employed or mixed with a WDM solution.

Figure 7.3 Passive WDM (O-RAN.WG9.WDM.0-R003-v03.00 2022). *Source:* With permission of O-RAN Alliance.

7.3.1 Passive WDM

The passive WDM solution is based on the end-to-end all-passive method. To achieve a low-cost WDM transmission, as shown in Figure 7.3, the O-RU directly uses the fixed or tunable optical transceivers, connected to the passive multiplexer/demultiplexer. At the O-DU side, the passive multiplexer/demultiplexer performs wavelength multiplexing/demultiplexing, which realizes the one-to-one optical wavelength connection.

Because the equipment at the O-DU side is only with colored optical transceivers, disadvantages of passive WDM include lack of management channel, weak perception of fiber link fault, the difficulty of optical transceiver operation, and maintenance that depends on manual work. Basic management functions of optical channels, fiber link fault, and the optical transceiver can be achieved in O-RU and O-DU host systems.

The equipment at the O-RU side and O-DU side may contain tunable optical transceivers, with active communication channels that exist within the optical carrier channel but do not affect the traffic. These transceivers allow the O-DU to exchange health, command, and control status information without the use of a supervisory channel. Multi-Source Agreement (Smart Tunable Module) specification for interoperable transceivers is part of a working group to allow O-RU and O-DU passive architectures to perform maintenance tasks.

7.3.2 Active WDM

The active WDM solution uses active WDM equipment for electrical and/or optical layer multiplexing at both the remote station and the central office, as shown in Figure 7.4. This solution also reduces the number of fibers and can provide management functions between WDM equipment.

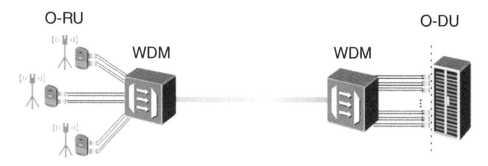

Figure 7.4 Active WDM (O-RAN.WG9.WDM.0-R003-v03.00 2022). *Source:* With permission of O-RAN Alliance.

Compared with the passive WDM, the cost of the active WDM scheme is about 4–6 times higher and is not conducive to large-scale deployment of 5G Fronthaul. Due to the remote location of WDM equipment at the O-RU side, available power sources and outdoor location may cause difficulty in the installation of this implementation.

7.3.3 Semiactive WDM

With the accelerated deployment of 5G networks, the Fronthaul network will have thousands of nodes creating a need for management of the capability of the network.

The proposed semiactive WDM schemes are illustrated in Figure 7.5.

This semiactive WDM solution is a simplification of the active WDM solution and an enhancement of the passive solution. Passive WDM at the remote O-RU side is not subject to power supply restrictions. The WDM equipment at the O-DU side of the central office is active, which can achieve monitoring, fault detection, and protection capabilities.

The active WDM equipment can send management requests to the O-RU and manage the WDM optical modules within the O-RU, including query and configuration. The optical modules within the O-RU can receive management requests from the active WDM equipment and then send the Operation, Administration and Maintenance (OAM) information of the optical modules

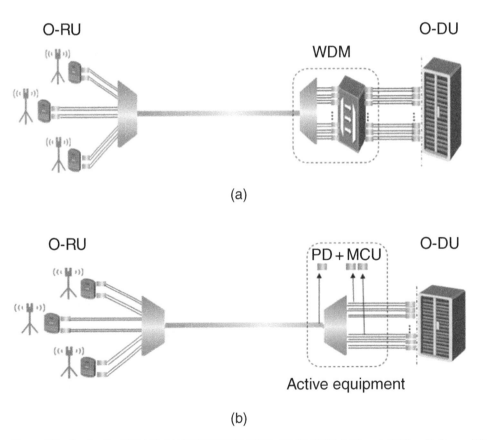

Figure 7.5 Semiactive WDM (O-RAN.WG9.WDM.0-R003-v03.00 2022). (a) Type I. (b) Type II. *Source:* With permission of O-RAN Alliance.

to the active WDM equipment, including the wavelength and output power of the transmitter. The optical modules in the O-RU and the O-DU can send the OAM information of optical modules to the active WDM equipment automatically or at regular time intervals once the optical modules are powered on. The WDM optical modules can add the OAM information with the service signals and transport them together in the same optical channel. The detection unit in the active WDM equipment can demodulate the OAM information, obtain the transmission performance of O-RU and O-DU, and then report it to the control system through the standard southbound interface. The semiactive WDM type I equipment should support query, configuration, and send OAM information. To further reduce system cost, the semiactive WDM type II equipment can perform simplified management, including sending the OAM information of optical modules to the active WDM equipment.

The semiactive WDM solution not only greatly reduces the pressure of optical fiber resources but also has cost advantages (compared with the active solution), management, and protection outside the optical transceiver and O-RU/O-DU host systems (compared with the passive solution). It helps operators build 5G Fronthaul networks with low cost, high bandwidth, and fast deployment.

7.4 Packet-Switched Solutions

This section provides an overview of alternatives to solve the connectivity scenarios between O-RU, O-DU, and O-CU leveraging on packet-switched technologies, as a summary of O-RAN specification in (ORAN.WG9.XPAAS.0-R003-v04.00 2022).

Packet switching devices on an operator's aggregation network can be deployed alone covering from the radio cell site up to the transport core location or can be combined with other technologies, for example, WDM as reported before, in a multilayer fashion (Doverspike et al. 2010). Such multilayer fashion is the common norm for internal segments of the network (e.g. backbone) in order to groom multiple sources of traffic and reach long distances, while toward the access, where a higher capillarity exists, a mix of approaches can be found. Here we will concentrate on the packet-switching layer itself, as the technology of interest, for satisfying the transport requirements for the Xhaul Transport, that is, at the Fronthaul, Midhaul, and Backhaul network segments. In general terms, we will refer to them as routers in this section.

O-RAN-disaggregated solutions more commonly will be deployed on brownfield scenarios, that is, coexisting with other mobile solutions (i.e. conventional 4G or 5G radio access networks) and even with other kinds of services (i.e. enterprise or fixed services), all using the same routers as transport infrastructure. This has implications in the sense that Xhaul traffic emerges as a new component to harmonize and integrate with the rest of traffic flows.

The Fronthaul traffic is formed by eCPRI or CPRI (encapsulated by the RoE protocol) frames, depending on the functional split considered in the radio part. In the case of O-RAN-based deployments, the split 7-2x advocates for eCPRI. However, packet-switched devices can be used for connecting CPRI flows (e.g. for split 8) and later convert them into eCPRI flows to be transported deeper into the network. The eCPRI (user data) traffic can be offered as Ethernet packets or using IP/UDP encapsulation, the former being more common. On the other hand, Midhaul and Backhaul portions of the traffic are typically encapsulated on IP data packets. Thus, the Xhaul transport solution requires handling these different components, with different requirements and distinct framing characteristics, for accomplishing the mission of aggregating all the flows from O-RU to O-CU.

7.4.1 Technology Overview

Packet switching is a very versatile traffic aggregation solution that permits very flexible deployments, facilitating the interworking of network elements (even in a multivendor interoperable approach) in a secure and redundant manner, and relaying standard protocols and encapsulation methods. Packet switching is an intrinsically efficient technology in terms of capacity consumption since it allows statistical multiplexing of the carried IP and Ethernet flows. In order to support prioritized traffic, the routers implement mechanisms of traffic prioritization, based on the marking of packets with different Quality of Service (QoS) values, as described later.

However, in some cases, the flows of interest could however require a more deterministic behavior, especially in the Fronthaul. Even in that case, packet-switched solutions incorporate specific technologies, which permit to enforce that behavior when requirements cannot be relaxed. This is the case of technologies like time-sensitive networking (TSN) (IEEE 802.1CM 2018) or FlexEthernet (OIF 2021) which implement specific mechanisms to guarantee reduced latency (and jitter) in the transmission (in the case of TSN, this is achieved by pre-emption mechanisms at the port queues, while in the case of FlexE, this is achieved by dedicating specific Ethernet frame slots which can be cross-connected at the MAC level).

Existing commercial routers implement a variety of different technology solutions which can be leveraged for accomplishing the Xhaul transport, and using one particular technology will depend on the specific deployment in each case (conditioned by existing services and devices, available infrastructure, operator best practices, etc.).

7.4.2 Deployment Patterns

Toward the access, and depending both on the availability of fiber infrastructure, the distances to cover (which limit essentially the achievable latency), and the geographical distribution of the sites (here determined by the presence of O-RU, O-DU, and O-CU), very different topologies can be present. Thus, it can be necessary to deploy devices conforming to rings, chains, or hub-and-spoke topologies.

The number of devices per aggregation area can impact the observed performance and cost of the network. Two main aspects can be considered. First, the impact on latency (and jitter), which, in some cases, such as the Fronthaul, is very limited and that has some physical, unavoidable dependencies like the delay due to the propagation of the light along the fibers connecting the routers (~5 µs/km), or the processing time of each device. Second, the amount of bandwidth necessary to provision to collect Fronthaul and Midhaul traffic from the covered geographical areas.

Thus, topologies involving a lower number of packet-based devices, like hub-and-spoke, preserve more latency budget available for increasing the network reach over larger distances (so enabling a higher degree of centralization of certain components, e.g. O-DU or O-CU) if compared to topologies with a higher number of routers, as in ring or chain topologies. The trade-off here is the availability of fiber since hub-and-spoke strategies require more available fibers than rings or chains.

Regarding bandwidth, topologies like rings and chains force the upgrade of the entire topology once one of the participant routers requires a capacity upgrade, for instance, due to the deployment of new O-RUs (or some other traffic source if the router is used for aggregating other services as well).

An additional and relevant aspect to consider is the impact on the clock signal distribution for synchronization purposes since each router in the path potentially contributes to increasing observed time errors (TEs).

7.4.3   Connectivity Service and Protocols

From the point of view of traffic delivery, it is possible to differentiate between the type of connectivity service and the protocols used for underlying forwarding of the traffic.

Regarding connectivity service types, both layer-2 (i.e. Ethernet-based) and layer-3 (i.e. IP-based) services can be supported in packet-switched networks. For layer-2 services, traditional L2VPNs or the next-generation Ethernet VPNs (EVPNs) are the alternatives. EVPN leverages BGP protocol extensions to disseminate MAC address information of every site participating in the VPN, as a difference with traditional L2VPNs, which determines the site reachability on MAC learning. Apart from that, EVPN introduces some other advantages such as traffic balancing. For the specific case of Xhaul transport, having layer-2 connectivity services is useful taking into account that Fronthaul traffic, such as eCPRI Radio over Ethernet (RoE), use Ethernet framing. Then, in scenarios where this kind of traffic could require transport among packet-switched devices, layer-2 connectivity can transparently do the job (always considering that the stringent requirements affecting Fronthaul traffic are satisfied). For layer-3 or routed services, conventional L3VPNs can be used, connecting Virtual Routing Functions (VRFs) instantiated in different routers per site in a conventional manner. This mechanism is commonly deployed today for Backhaul, and it is straightforward to extend it toward the Midhaul.

The aforementioned connectivity services can be carried on top of different forwarding protocols. With respect to the underlying protocols for traffic forwarding, the majority of current networks base their forwarding capabilities on multiprotocol label switching (MPLS) encapsulation. The MPLS data plane is based on the fast switching of labels at packet headers. The population of labels to be used to reach different destinations in the network can be performed by different mechanisms supported by additional control-signaling protocols. Since the purpose of this book is not to detail the specifics of a particular solution, the reader is suggested to look at (Minei and Lucek 2010) or (Sánchez-Monge and Szarkowicz 2015) for a deeper insight. An interesting capability enabled by MPLS is the possibility of performing traffic engineering (TE) by determining concrete Label Switch Paths (LSPs) that can be planned for different purposes, such as optimizing the traffic delivery (e.g. in terms of latency) or the consumption of resources (e.g. bandwidth usage). In traditional MPLS, the way of determining those LSPs involves the signaling and interchange of reachability information between routers in the network, and, in some cases, the participation of centralized control elements (e.g. the path computation element, PCE). In conventional MPLS, intermediate transit nodes need to maintain path information.

There is however a recent evolution with the purpose of alleviating the amount of signaling tasks and protocols in the network which essentially advocates for the introduction of the path information only at the ingress node, following a source routing model. This is known as segment routing MPLS (SR-MPLS). With that purpose, SR-MPLS divides the packet forwarding path into different segments (either prefix, node, or adjacency segments), allocating segment identifiers (SIDs) to them, with that segment information being encapsulated into packets at the ingress of the path. The SIDs structurally follow the format of an MPLS label. That SIDs are propagated in a lightweight manner compared to conventional MPLS, not requiring state information in intermediate routers, as well. SR-MPLS also allows the exploitation of TE mechanisms for optimization purposes.

Relying on the same idea of source routing, segment routing IPv6 (SRv6) appears as another alternative for underlying delivery. SRv6 is implemented on the IPv6 forwarding plane, then not requiring MPLS labels for accomplishing the data packet forwarding. In the case of SRv6, the SIDs are contained in the IPv6 header as segment routing headers (SRHs). From a control plane perspective, SRv6 can also leverage a centralized PCE for the calculation of paths.

For an up-to-date view of SR-MPLS and SRv6, the reader is referred to (Ventre et al. 2021).

7.4.4 Quality of Service (QoS)

The possibility of marking the packets at Ethernet, MPLS or IP levels with different QoS values allows to differentiate flows according to their importance and apply different priorities which are reflected in the way those packets are treated at the port queues in the routers. Thus, the higher priority packets are delivered first and lower priority packets could be discarded (i.e. in case a low-priority packet arrives and the buffer space allocated for such low-priority packets become full).

QoS is a traditional mechanism deployed in operational networks for handling a variety of traffic flows. Usually, the high-priority flows have been a few and determined by the criticality of the carried service, e.g. voice communication. With the support of multiple services in the network, the strategy of planning the QoS markings and the dimensioning of the corresponding buffers are becoming more difficult. This difficulty is increased now with the consideration of Xhaul since some of the traffic flows, such as the Fronthaul ones, require high-priority treatment to satisfy the stringent communication requirements, but with the special characteristic of demanding very high bandwidth. This fact implies that the sizing of high-priority queues in routers has to change from the previous practices, and in consequence, careful planning is required, considering not only Xhaul part but the rest of services aggregated in the same network segment (i.e. conventional 4G or 5G mobile traffic, enterprise, or fixed services) as well.

7.5 Management and Control Interface

Operational service provider transport networks are usually built leveraging multivendor solutions not only across layers (i.e. IP and optical) but even in a single layer. Traditional management of these transport networks has leveraged proprietary network management systems (NMSs) per vendor technology, usually integrated through a common umbrella management system. This way of proceeding implies a continuous effort of integration since proprietary NMSs typically expose nonstandard interfaces, which require parameter and procedure adaptations when new features or functionalities become available.

With the advent of network programmability at the pace marked by the software-defined networking (SDN) trend, the industry now is developing common standards and data models for control and management purposes. The final objective, from operators' perspective, is to target a common manner of operating the transport network, independently of the particular device being managed.

Several standardization development organizations (SDOs) are working on this area, with prevalence on IETF (for IP and Ethernet technologies) and ONF (for optics) to which Xhaul transport refers. This section provides an overview of the current state-of-the-art in this respect, as well as proposes some hints for further integration with O-RAN management layer.

7.5.1 Control and Management Architecture

Operators are pursuing the definition of scalable architectures of control for their transport networks. In this respect, a generic schema can be represented by the Open Transport SDN Architecture (Telecom Infra Project 2020) promoted by the Telecom Infra Project (TIP). It proposes a hierarchical architecture, as depicted in Figure 7.6.

Transport services, in general, span several technology domains (namely IP/MPLS, optical or microwave) and their corresponding resources. Because of the particularities of each technology,

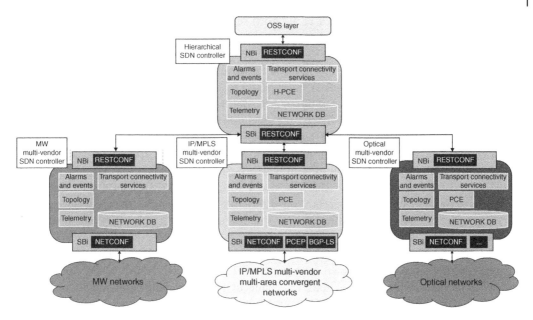

Figure 7.6 TIP open transport SDN architecture.

not only in terms of physical characteristics but also related to protocols and supported features, it seems advantageous to deploy specific technology domain controllers, being able to fully handle the resources underneath. This way of proceeding allows to better specify the responsibility boundaries for each of the per-domain controllers, while at the same time providing higher scalability of the overall solution. Finally, it permits the selection of the best-of-breed control solution per technology domain.

The hierarchical SDN controller laying on top of the per-domain controllers acts as a single entry point for any connectivity request. For doing that, the hierarchical controller supports through its northbound interface (NBI) generic-service application programming interfaces (APIs) and an abstracted topological view of the transport resources available. Those APIs and abstractions could differ depending on the client demanding transport services (that is, the operator OSS, internal service orchestrators, or specific higher-level service controllers, such as the one from O-RAN).

The described three-tier solution is aligned with other initiatives in the industry, such as the IETF Framework for Service Automation (Wu et al. 2021). Operators like Telefonica follow this same design. For reference, in (Contreras et al. 2019), the so-called iFUSION control architecture is described in detail as the control and management target architecture being deployed on Telefonica operations.

Since the transport network is common and unique for the multiple services supported by carriers, the expectation is that different consumers of the transport network interact with these control capabilities when requesting connectivity services. This also applies to the Xhaul. In consequence, it can be expected an interaction between the control capabilities of O-RAN and the ones of the transport network (i.e. the hierarchical SDN controller of Figure 7.6) for the dynamic provisioning of the connections between O-RU, O-DU, and O-CU, when needed.

In terms of NBI, it can be expected in the hierarchical SDN controller at least the support of virtual private networks (VPNs) service models as defined by IETF. Those refer to layer-2 VPN

(Wen et al. 2018) or layer-3 VPN (Wu et al. 2018) service models. These are also the kind of connectivity services expected in current O-RAN specifications, e.g. (ORAN.WG9.XPAAS. R003-0-v04.00 2022).

Regarding the interaction with the network elements of the distinct technologies under consideration (IP, microwave, and optics), the objective is also to leverage common data models for each of them. This implies the usage of different models as summarized in (O-RAN.WG9.XTRP-MGT.0-R003-v05.00 2022), which overviews the outcomes from different SDOs in this respect.

For IP and Ethernet transport, IETF-based models can be considered, for both network and device configuration and management. At the network level, the objective is to translate service model information into the configuration of network-wise parameters, such could be the case of the virtual routing functions (VRFs) associated with an L3VPN. At the device level, the purpose is to translate particular aspects of the device to support the network model before, such as QoS parameters. For a description of how the different models imbricate, please refer to (Wu et al. 2021).

In the case of optics, a suitable model is defined by ONF for configuring the optical domain at the network level. That model, known as Transport API (TAPI) (ONF Transport API), provides a number of services such as topology retrieval, connectivity request, path computation, or virtual network, that allows the configuration of an optical domain. Going to the device level, some other models can be leveraged, such as OpenConfig (OpenConfig), with the purpose of controlling specific parts and parameters of an optical device.

Finally, for wireless transport or microwave technology, another ONF model, the TR-532 (ONF TR-532 v2.0), allows the automated configuration of microwave devices.

All the aforementioned models can be in principle used on Fronthaul, Midhaul, and Backhaul scenarios implemented with the variety of technologies here described, in a coordinated manner, through the hierarchical control architecture described.

7.5.2 Interaction with O-RAN Management

In the current O-RAN architecture, as defined in (O-RAN.WG1.O-RAN-Architecture-Description-REV003-v08.00 2022), the service management and orchestration (SMO) does not define any interface between O-RAN and the equivalent control components at the transport network. Thus, O-RAN is conceived as an overlay on top of the transport network without direct interaction among them.

However, in order to exploit the advantages of a fully programmable network, it can be expected an interaction at the control level, permitting not only the invocation of simple connectivity services, as the mentioned VPNs, but also future advanced services as devised for the case of network slicing.

Considering the generic transport SDN architecture here, it can be assumed the SMO is a new "client" of the transport network, then consuming the NBI offered by the hierarchical SDN controller in a similar way as could happen with other consumers of connectivity services. Through such an interface, the SMO can request connectivity services adapted to the needs of each particular deployment, enabling an integrated and automated configuration of the O-RAN and Xhaul capabilities end-to-end. Therefore, it may make sense to add the transport network into the overall O-RAN architecture as depicted in Figure 7.7.

The functional definition of such an interface is outside the scope of O-RAN since, as mentioned, O-RAN represents another consumer of connectivity services of a carrier network. Then such an

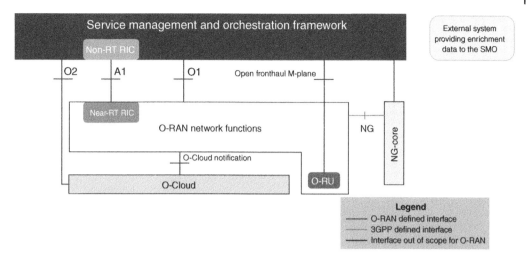

Figure 7.7 Overall O-RAN Architecture (*Source:* O-RAN.WG1.O-RAN Architecture Description-REV003-v08.00 (2022))

interface should be generic, even though offering sufficient capabilities as needed for O-RAN services. Here it is assumed that such an interface is compliant with the NBI of the hierarchical SDN controller described before.

7.6 Synchronization Solutions

Meeting the emerging 5G synchronization requirements stated in Section 7.2.4 demands new synchronization architectures and solutions. In contrast to previous architectures, the Fronthaul network takes a more central role in the analysis and design of this new architecture. Furthermore, stringent TE limits constrain the number and type of clocks within the synchronization chain.

7.6.1 Synchronization Baseline

Following synchronization topologies are possible within Xhaul networks (Figure 7.8):

- **LLS-C1:** O-RUs receive their synchronization reference from the O-DUs.
- **LLS-C2:** O-DUs are still in the synchronization chain to the O-RUs, but there are intermediate clocks.
- **LLS-C3:** Both O-DUs and O-RUs receive their reference from the Fronthaul network.
- **LLS-C4:** O-RUs receive their synchronization reference locally.

This figure assumes the use of full timing support (ITU-T G.8275.1) for PTP in Fronthaul networks. The deployment of partial timing support (ITU-T G.8275.2) is permissible but demands that system operators ensure that the network components meet the frequency and phase error requirements indicated in the synchronization requirement section.

The use of synchronous Ethernet according to ITU-T G.781 (Option 1 Quality Level) and ITU-T G.8264 is strongly recommended. Synchronous Ethernet is a strong example of a physical layer frequency signal (PLFS). Other PLFS implementations are being studied at the time.

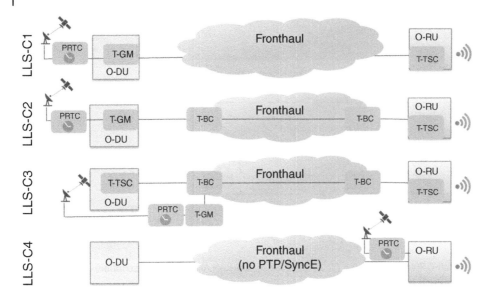

Figure 7.8 Synchronization topologies.

7.6.2 Synchronization Accuracy and Limits

While the TAE metrics provide information on measurements over the air, the TE is used to validate the synchronization accuracy over the UNI in Xhaul networks (Figure 7.2). The TE (t) identified the difference in time/phase from a reference which is typically represented by a global navigation satellite system (GNSS). The TE is further characterized by the following metrics:

- **Max|TE|:** The maximum absolute value of the TE function TE(t).
- **Constant Time Error, cTE:** The mean value of the TE function, TE(t), over a measurement period.
- **Dynamic Time Error, dTE(t):** The change in the TE function, TE(t), over a measurement period.

Beyond TAE, the air interface frequency error represents the other main synchronization metrics. 3GPP specifies ± 50 ppb as the limit for this metric.

Translating 3GPP air interface time alignment and frequency error metrics, O-RAN CUS specification [WG4-CUS] provides limits and budgets for time and frequency error over Fronthaul networks UNI. The budgets and limits are dependent on the following factors. For examples, please refer to tables 11.3.2.1-1 through 11.3.2.3-1 in (O-RAN-WG4.CUS.0-v11.00 2022).

- **Synchronization configuration mode:** LLS-C1, LLS-C2, LLS-C3, LLS-C4
- **Synchronization category:** A+, A, B, C
- **O-RU category:** regular or enhanced
- **O-DU master class:** A, B

7.7 Testing

Testing of Xhaul transport alternatives is an essential task in order not only to ensure the validity of the existing solutions for Xhaul service assurance and satisfaction of requirements but also to learn from the practice and obtain technical deployment guidelines before moving transport solutions into production, on the field.

Table 7.4 Example of Fronthaul and Midhaul dimensioning for scenarios in [O-RAN.WG9.XTRP-REQ-v01.00 2020].

	Fronthaul (O-RU ↔ O-DU)		Midhaul (O-DU ↔ O-CU)	
	Peak capacity (Gbps)	Interfaces	Peak capacity (Gbps)	Interfaces
Small site FR1 only	7.5	1×10GE	2	1×10GE
Small site FR2 only	14.57	1×25GE	3.7	1×10GE
Medium site FR1 + FR2	110	6×25GE 3×10GE	28.4	1×100GE
Large site FR1 + FR2	264.8	18×25GE	68.9	1×100GE

In carrier networks, it is almost impossible to find a single architectural pattern, thus running both functional and performance testing campaigns, which helps revisit scenarios and identify potential limitations that can happen to some particular operators but not to others.

With that aim, O-RAN WG9 has released initial testing specifications for Xhaul (O-RAN.WG9. XTRP-TST-v01.01 2021) covering the technologies here described in a wide sense, including synchronization aspects. This specification collects more than 80 tests in its current version for a complete assessment of the technologies of interest.

At the time of testing, it is also necessary to characterize the scenario under test with realistic assumptions. There are several dimensions of testing to cover, as follows:

- Performance indicators or service-level objectives (SLOs), such as throughput, latency, jitter, etc., in line with the expectations for Fronthaul, Midhaul, and Backhaul segments.
- Technologies used for Xhaul transport, such as different kinds of flow framing (eCPRI, IP, etc.), distinct basic technologies (optics, packet-switched, microwave, etc.), and data plane alternatives (WDM, conventional IP and Ethernet, TSN, FlexE, etc.).
- Protocols and encapsulations for service delivery and data forwarding (L2VPN, L3VPN, MPLS, SRv6, etc.).
- Possible topology structures (ring, chains, hub-and-spoke, etc.).
- Co-existence of other traffic contributions from other services in the network (conventional 4G and 5G, fixed, enterprise, etc.).
- Different connectivity schemas (e.g. O-DU in the cell site or remote).

It is also interesting to emulate deployment scenarios that could mimic real situations. This implies to dimension the necessary connectivity as could be found in a real deployment. For instance, Table 7.4 represents the number and type of interfaces that could be needed for satisfying exemplary scenarios described in (O-RAN.WG9.XTRP-REQ-v01.00 2020).

At the time of writing, several Xhaul testing campaigns have been performed as part of O-RAN plugfest events, serving as validation points of test specifications as well as a way of providing continuous feedback for up-to-date maintenance of test descriptions.

7.8 Conclusion

The introduction of smart phones and new 5G services has increased the role of transport networks that interconnect a large number of cell sites and radio elements needed to deliver the required bandwidth and latency requirements. This chapter started by listing some critical requirements such as one-way latency, frame loss ratio, and bandwidth requirements. These requirements can be met by physical layer, packet-switched and synchronization, and management interface technologies that are being enhanced for O-RAN. Finally, a set of test cases were introduced to validate the functions and performance of open Xhaul transport networks.

Bibliography

Contreras, L.M., González, Ó., López, V., et al. (2019). *iFUSION: Standards-based SDN Architecture for Carrier Transport Network*, IEEE Conference on Standards for Communications and Networking (CSCN), Granada, Spain.

Doverspike, R.D., Ramakrishnan, K.K., and Chase, C. (2010). Structural Overview of Commercial Long Distance IP Networks. In: *Chapter 2 in Guide to Reliable Internet Services and Applications*, 1ste (ed. C. Kalmanek, S. Misra, and R. Yang). Springer.

IEEE 802.1CM (2018). *Time-Sensitive Networking for Fronthaul.*

Minei, I. and Lucek, J. (2010). *MPLS-Enabled Applications: Emerging Developments and New Technologies*, 3e. Wiley.

OIF (2021). *FlexE 2.2 Implementation Agreement, IA # OIF-FLEXE-02.2.*

ONF TR-532 v2.0. *Microwave Information Model.*

ONF Transport API. https://github.com/OpenNetworkingFoundation/TAPI.WG1

OpenConfig. https://openconfig.net/projects/models.

O-RAN.WG1.O-RAN-Architecture-Description-REV003-v08.00 (2022). *O-RAN Architecture Description.*

O-RAN-WG4.CUS.0-v11.00 (2022). *O-RAN Fronthaul Working Group Control, User and Synchronization Plane Specification.*

O-RAN.WG9.XTRP-MGT.0-R003-v05.00 (2022). *Management Interfaces for Transport Network Elements.*

O-RAN.WG9.XTRP-REQ-v01.00 (2020). *Xhaul Transport Requirements.*

O-RAN.WG9.XTRP-TST-v01.01 (2021). *Xhaul Transport Testing.*

O-RAN.WG9.WDM.0-R003-v03.00 (2022). *WDM-based Fronthaul Networks.*

ORAN.WG9.XPAAS.0-R003-v04.00 (2022). *Xhaul Packet Switched Architectures and Solutions.*

Sánchez-Monge, A. and Szarkowicz, K.G. (2015). *MPLS in the SDN Era*, O'Reilly.

Telecom Infra Project (2020). *Open Transport SDN Architecture*, Online.

Ventre, P.L., Salsano, S., Polverini, M., et al. (2021). *Segment Routing: A Comprehensive Survey of Research Activities, Standardization Efforts, and Implementation Results*, IEEE Communications Surveys & Tutorials, Vol. 23, No. 1.

Wen, B., Fioccola, G., Xie, C., et al. (2018). *Data Model for Layer 2 Virtual Private Network (L2VPN) Service Delivery, RFC 8466.*

Wu, Q., Litkowski, S., Tomotaki, L., et al. (2018). *YANG Data Model for L3VPN Service Delivery, RFC 8299.*

Wu, Q., Boucadair, M., Lopez, D., et al. (2021). *A Framework for Automating Service and Network Management with YANG, RFC 8969.*

8

O-RAN Security

Amy Zwarico

AT&T, Dallas, TX, USA

8.1 Introduction

The open RAN architecture is a fundamental change from traditional proprietary, hardware-based RAN deployments. It disaggregates and opens the traditional 3GPP RAN architecture into well-defined components with standardized interfaces and introduces non- and near-RT intelligent controllers that enable operators to embed custom intelligence into the RAN. This architecture increases the threat surface from that of the traditional RAN by adding new functions as well as more interfaces. At the same time, however, the open nature of the disaggregation introduces new opportunities to strengthen the security of the radio access network.

In this chapter, we delve into the evolving security of O-RAN. We begin with a review of the zero trust principles driving O-RAN security controls. The next section of the chapter is devoted to an analysis of the threats to O-RAN deployments. The third section describes the security controls supported by O-RAN components by design. We end with security recommendations for operators deploying O-RAN.

8.2 Zero Trust Principles

O-RAN security controls are designed based on the principle of zero trust and zero trust architectures described in the NIST Zero Trust Architecture (Rose et al. 2020). Zero trust is the term for an evolving set of cybersecurity paradigms that move defenses from static, network-based perimeters to dynamic controls that focus on protecting users, assets, and resources in a network viewed as compromised. Zero trust is a natural evolution of security resulting from a more sophisticated understanding of threats and how to prevent, detect, and mitigate. Zero trust is about recognizing that bad actors will always be on your network, so a strong perimeter is not enough. At a bare minimum, each resource must be protected by a policy enforcement point that authenticates each session to a resource, encrypts the session to prevent eavesdropping, and enforces policies that ensure least privilege access to resources. In addition, zero trust uses information from the continuous monitoring of the environment to make dynamic access decisions. The information is gathered from the network, the resources, assets, and even from external sources to detect changes that may

Open RAN: The Definitive Guide, First Edition. Edited by Ian C. Wong, Aditya Chopra, Sridhar Rajagopal, and Rittwik Jana.
© 2024 The Institute of Electrical and Electronics Engineers, Inc. Published 2024 by John Wiley & Sons, Inc.

indicate the presence of a bad actor, a change in the security posture of an asset or resource, new zero-day attacks, and even new standards.

Remembering the NIST basic tenets of zero trust (Rose et al. 2020) will help in understanding the security controls designed into O-RAN and additional controls, especially monitoring and identity management, that a mobile network operator (MNO) will need to securely operate an O-RAN deployment.

1. All data sources and computing services are considered resources.
2. All communication is secured regardless of network location.
3. Access to individual enterprise resources is granted on a per-session basis.
4. Access to resources is determined by dynamic policy – including the observable state of client identity, application/service, and the requesting asset – and may include other behavioral and environmental attributes.
5. The enterprise monitors and measures the integrity and security posture of all owned and associated assets.
6. All resource authentication and authorization are dynamic and strictly enforced before access is allowed.
7. The enterprise collects as much information as possible about the current state of assets, network infrastructure, and communications and uses it to improve its security posture.

8.3 Threats to O-RAN

This section provides a deep dive into the existing and emerging threats against O-RAN and its underlying virtualization platforms. The threats against a 5G RAN have been studied by the 3GPP, O-RAN Alliance [SFG Threat Analysis], and the Enduring Security Framework (ESF) (NSA and CISA Part I 2021; NSA and CISA Part II 2021; NSA and CISA Part III 2021; CISA NSA 2021). The Cloud Security Alliance (CSA) and Cloud Infrastructure Telco Taskforce (CNTT) have written extensively about virtualization threats.

8.3.1 Stakeholders

The threats to O-RAN are of particular interest to anyone involved in the implementation, management, operation, and maintenance of the O-RAN system. The list is long and includes MNOs, orchestrators, hardware/network vendors, hardware/network administrators, network function (NF) vendors, NF administrators, virtualization/containerization hardware infrastructure providers, virtualization/containerization hardware infrastructure administrators, virtualization/containerization software infrastructure providers, virtualization/containerization software infrastructure administrators, system integrators, system testers, identity administrators, role-based access control (RBAC) administrators, system administrators, and public key infrastructure (PKI) administrators (O-RAN ALLIANCE O-RAN.SFG.O-RAN-Threat-Model-v03.00 2022). It should be noted that zero trust assumes that a bad actor could assume any of these roles.

8.3.2 Threat Surface and Threat Actors

The O-RAN threat surface includes the O-RAN-specific functions and interfaces, the newly introduced lower layer split (LLS) architecture, virtualization, and supply chain (O-RAN ALLIANCE O-RAN.SFG.O-RAN-Threat-Model-v03.00 2022).

The threat actors include cyber-criminals, malicious insiders, hacktivists, cyber-terrorists, script kiddies, and nation-state actors. Malicious insiders are assumed to have authorized system access to at least one O-RAN asset or another system that would provide access to an O-RAN asset.

It is assumed that an attack can originate outside of O-RAN, for example, from an infected IoT device or by an attack through an interface, or from inside an O-RAN component, such as the SMO or a malicious xApp. These attacks can compromise the availability of mobility network components and the confidentiality and integrity of the data handled by O-RAN (O-RAN ALLIANCE O-RAN.SFG.O-RAN-Threat-Model-v03.00 2022). They are divided into the O-RAN-specific vulnerabilities:

- Unauthorized access to O-DU, O-CU-CP, O-CU-UP, and RU to degrade RAN performance or execute broader network attack (Availability)
- Unprotected synchronization and control plane traffic on Open Fronthaul Interface (Integrity and Availability)
- Unauthorized disabling of over-the-air ciphers for eavesdropping (Confidentiality)
- Near-RT RIC conflicts with O-gNB (Availability)
- x/rApps conflicts (Availability)
- x/rApps access to network and subscriber data (Confidentiality)
- Unprotected management interface (Confidentiality, Integrity, and Availability)
- Injection of uplink (UL) or downlink (DL) messages (Availability)

and general vulnerabilities:

- Decoupling of functions from hardware root of trust (Integrity)
- Problems with the software trust chain (Integrity)
- Exposure to known exploits in open source code (Confidentiality, Integrity, and Availability)
- Misconfiguration, poor isolation, or insufficient access management in the O-Cloud platform (Confidentiality, Integrity, and Availability)

In the following sections, the threats to O-RAN are described in more detail. Additional information can be found in (O-RAN ALLIANCE O-RAN.SFG.O-RAN-Threat-Model-v03.00 2022).

8.3.3 Overall Threats

Without properly designed and configured security controls, an attacker could successfully launch any number of attacks. Missing or misconfigured security controls can lead to weak or missing authentication, missing access control, undocumented open ports, weak protections on data on the O-RAN components, and weak protections on log files. Exploiting weak authentication on an interface, an attacker could access configuration data on an O-RAN component. An attacker could use an undocumented open port to exfiltrate data from an O-RAN component.

8.3.4 Threats Against the Lower Layer Split (LLS) Architecture and Open Fronthaul Interface

The O-RAN LLS architecture separates baseband unit (BBU) functionality into the O-CU-CP, O-CU-UP, and O-DU, increasing the threat surface to include these disaggregated BBU functions, the Open Fronthaul interfaces (M/C/U/S-planes), the software implementing these functions, the virtualization technologies, and the underlying hardware platforms. In this section, we describe the threats against the Open Fronthaul and O-RAN functions. Threats against virtualization technologies and hardware will be discussed later as they affect all O-RAN components.

If an attacker gains unauthorized access to the Open Fronthaul Ethernet L1 physical layer interface, he can attach a rogue device to the Ethernet L1 interface that floods the L1 interface with bogus network traffic resulting in disruption or degradation in the performance of the authorized network elements on the Open Fronthaul interface. The rogue device could also degrade or disrupt the Open Fronthaul interface by sending bogus L2 messages. With physical access, the attacker could damage an Ethernet port connection or cut the physical interface (coaxial cable, twisted pair, or optical fiber).

An attacker could launch S-plane attacks that disrupt the timing between the O-DU and O-RU. For example, by hijacking the S-plane, the attacker could launch a DoS attack against a Master clock. The attacker could impersonate or spoof the master clock within the precision time protocol (PTP) network by issuing a fake ANNOUNCE message. With physical access, an attacker could add a rogue PTP instance that acts as a Grand Master. Finally, the attacker could remove or manipulate PTP timing packets, degrading performance.

A successful man-in-the-middle (MITM) attack could be launched against an improperly secured M-plane or U-plane.

Finally, the attacker could spoof the DL or UL C-plane messages.

8.3.5 Threats Against O-RU

An attacker could deploy a rogue O-RU and fool either an O-DU or UE into attaching to the rogue O-RU instead of a legitimate O-RU. Rogue or false O-RUs, also known as SUPI/5G-GUTI catchers, can retrieve the subscriber's identity by sniffing the unencrypted traffic over the air when the UE is attached to the rogue O-RU. This allows the attacker to intercept the subscriber's identity and track the subscriber via their UE.

8.3.6 Threats Against Near- and Non-Real-Time RICs, xApps, and rApps

As discussed in previous chapters, the near- and non-RT RICs are new constructs in the O-RAN architecture that allow an operator to add intelligence and machine learning (ML) to control the network through the inclusion of xApps and rApps. A malicious xApp can be used to exploit UE identification, track UE location, and change UE priority. An attacker could penetrate the non-RT RIC and send messages to the SMO via the R1 interface or to the near-RT RIC via the A1 interface to cause a denial or degradation in service or to obtain UE data.

x/rApps are typically AI or ML applications used to analyze the state of the network and drive other actions, such as network tuning. A malicious xApp could provide incorrect information that causes degradation in network performance. Conflicting x/rApps unintentionally or maliciously impact an O-RAN deployment by repeatedly sending conflicting messages creating a race condition in the network. A malicious xApp could compromise xApp isolation, obtaining potentially sensitive information about UE or network behavior. An attacker could train an x/rApp with bad data leading (data poisoning) to faulty information about network performance. They could alter an ML model in ways that lead to network performance degradation or extract sensitive or confidential data.

8.3.7 Threats Against Physical Network Functions (PNFs)

An O-RAN deployment may be a mix of PNFs and VNF/CNF. An attacker could take advantage of weak legacy security controls on a PNF to perform attacks against the VNFs/CNFs in the network.

8.3.8   Threats Against SMO

An attack against the SMO could be devastating because the SMO is responsible for managing and orchestrating O-RAN. An attacker could exploit improper or missing authentication and authorization controls to gain control over the O-RAN deployment. With this control, the attacker could access sensitive information about the O-RAN components and UE (surveillance), degrade performance, consume all compute and storage capabilities in the underlying compute, damage expensive radio units, disable RIC functionality, install malicious x/rApps in the network, and launch attacks against the mobility core.

8.3.9   Threats Against O-Cloud

The BBU functions traditionally delivered as tightly coupled hardware and software from a single vendor and managed by an MNO within their controlled environment will now be delivered as software packages deployed to virtualized and potentially shared compute infrastructure. As the use of public clouds increases, the hardware and virtualization infrastructure will be managed by a third party and may be shared by multiple unrelated companies. The attack surface now includes the virtualization platform such as the hypervisor, container orchestrator, container engine, and the repository storing software images.

A single hypervisor or container orchestrator may run unrelated VMs/containers, each managed by different teams, and with different sensitivity levels. Misconfigurations in the VMs, containers, or virtual networks can allow a malicious application to degrade the execution, access the data, or eavesdrop on the virtual network of another VNF/CNF running on the same O-Cloud. A malicious application running on the same virtualization infrastructure could launch an attack on the hypervisor, storage, memory, or network of the underlying compute environment. Finally, a malicious application running with elevated privileges could compromise the isolation of the O-RAN components sharing the virtualization platform.

VNF/CNF images are pulled from an MNO repository or registry as the O-RAN is deployed and scaled up to respond to increased load. While in the repository or registry, an attack may tamper with the images or access embedded secrets.

3GPP, NIST, ENISA, CSA, and CISA, among others, have in-depth studies of virtualization and containerization security.

8.3.10   Threats to the Supply Chain

Supply chain risk refers to efforts by threat actors to exploit information and communications technologies (ICTs) and their related supply chains for purposes of espionage, sabotage, foreign interference, and criminal activity (CISA NSA 2021). The use of open-source code has increased dramatically over the past decade (see https://www.sonatype.com/resources/white-paper-2021-state-of-the-software-supply-chain-report-2021), and O-RAN software will contain increasing amounts of open source. O-RAN software can include packages with known vulnerabilities, malicious code known only to the package developer, and packages from untrusted sources. The vulnerabilities, malicious code, and other code weaknesses can be exploited to exfiltrate data, launch injection attacks, degrade performance, and cause remote code execution. Fixing problems in the code will take time, while testing and deploying those fixes in a running network requires extensive planning. Vendors and MNOs may find themselves in a constant state of patching, while new feature development receives lower priority.

8.3.11 Physical Threats

An MNO will have to protect against physical threats. An attacker with physical access to O-RAN components or the fronthaul cable network could access sensitive data, physically damage the computer or cable network, install sniffers on the network, or disrupt power. Natural and man-made disasters can also disrupt O-RAN operation.

8.3.12 Threats Against 5G Radio Networks

5G radios, like previous generation radios, can be jammed and spoofed. If encryption is misconfigured in the radio, then an attacker can obtain sensitive data transmitted between the radios and UE or between the radios and O-RAN components. A denial of service attack against a cognitive radio, a technology designed to enhance spectrum utilization, can occur if the attacker injects interference or tricks an unlicensed secondary user into believing that the unoccupied spectrum is being used. For more on cognitive radio attacks, see https://arxiv.org/pdf/1603.01315.pdf.

8.3.13 Threats to Standards Development

O-RAN security is driven by the O-RAN Alliance security standards. The ESF has identified two potential risks that can affect security. First, nation-states may attempt to exert undue influence on standards that benefit their proprietary, potentially untrusted, technologies, and limit customers' choices to use other equipment or software. The second risk is that optional standards may not be provided by vendors or implemented by an MNO (CISA NSA 2021). Such omissions will leave an O-RAN deployment susceptible to attack.

8.4 Protecting O-RAN

With an understanding of the potential threats against an O-RAN deployment, we now turn to the security controls used to thwart the all-too-real threats. These security controls are defined by the O-RAN Alliance in (O-RAN ALLIANCE O-RAN.SFG.O-RAN-Security-Requirements-Specifications-v03.00 2022; O-RAN ALLIANCE O-RAN.SFG.Security-Protocols-Specifications-v03.00 2022). We begin by describing the controls used to protect the O-RAN-defined interfaces. We then describe controls that need to be in place to protect data assets at rest. The following sections discuss security for the O-Cloud and for the virtualization techniques used in O-RAN implementations. We end with a description of O-RAN software security.

8.4.1 Securing the O-RAN-Defined Interfaces

O-RAN defines a set of interfaces that are used for management and interaction between components. Securing these interfaces by design is the first step in securing O-RAN. The goal is to integrate authenticity, integrity, confidentiality, and least privilege security controls into each interface specification.

Authenticity ensures that the identities of the two endpoints of the interface have been reliably verified.
Integrity ensures that the message has not been altered during transmission.

Confidentiality ensures that no unauthorized user (human or machine) can see the contents of the message during transmission.

Least privilege prevents the receiving endpoint from executing functionality on behalf of the sending endpoint that it is unauthorized to perform.

O-RAN has adopted widely used security protocols to protect both the O-RAN-defined interfaces as well as the 3GPP interfaces. The protocols include TLS 1.2, 1.3 (IETF RFC 5246 2008; IETF RFC 8446 2018), HTTP over TLS (IETF RFC 2818 2000), IPSec (IETF RFC 4303 2005; 3GPP 33.210), SSHv2 (IETF RFC 4252 2006), FTPS, and OAuth 2.0 (IETF RFC 6749 2012). O-RAN requires that current versions of the protocols be used with secure ciphers (encryption algorithms that are not broken). Using standard protocols provides security benefits because these protocols and their implementations are widely tested, extensively studied, and regularly updated as new features are required or vulnerabilities discovered. Both development and operations organizations have expertise in integrating and running these protocols. Finally, these protocols typically integrate easily with other security platforms, such as identity life cycle management. Current details about the security protocols are maintained by the O-RAN Alliance in the O-RAN Security Protocols Specifications (O-RAN ALLIANCE O-RAN.SFG.Security-Protocols-Specifications-v03.00 2022).

8.4.1.1 A1

The A1 interface between the non-RT RIC and near-RT RIC implements authenticity of both endpoints, confidentiality, and message integrity, and prevents replay attacks using TLS 1.2 or 1.3 configured to use strong ciphers only and to require both client and server authentication (mutual authentication or two-way TLS).

8.4.1.2 O1

The O1 interface (O-RAN ALLIANCE O-RAN.WG10.O1-Interface.0-v07.00 2022), used by the SMO to manage the O-eNB, non- and RT RICs, O-CU-UP, O-CU-UP, and O-DU, is implemented using the NETCONF protocol (IETF RFC 6241 2011). It leverages TLS 1.2 and 1.3 to implement the authenticity of both endpoints, confidentiality, and message integrity and prevents replay attacks. Like the A1 interface, TLS is configured with strong ciphers only and requires both client and server authentication (mutual authentication or two-way TLS). Alternatively, SSHv2 may be used to secure the O1 transport. O-RAN recommends the use of SSH keys for authentication, rather than passwords. As with TLS, O-RAN requires the use of strong ciphers only with SSH.

All O1 functions in an O-RAN deployment may be considered sensitive or vulnerable. Write operations (e.g. edit-config) to these functions without proper protection can have a negative effect on network operations. Unrestricted read operations can reveal sensitive information to unauthorized users. O-RAN uses the Network Configuration Access Control Model (NACM) (IETF RFC 8341 2018) to provide the means to restrict access for users to a preconfigured subset of all available NETCONF protocol operations and content. NACM further enables operators to integrate both authentication and authorization with a centralized access management platform, protect the running configuration from unauthorized modification and deletion, and support change management procedures for updating both NF configurations and changes to the NACM instance on an NF.

O-RAN configures NACM to be on by default. It permits read for authenticated users by default and disables both write and execute by default. Furthermore, O-RAN allows NACM to retrieve user role assignments from a centralized identity management system, such as Lightweight Directory Access Protocol (LDAP) or Terminal Access Controller Access Control System (TACACS).

The O-RAN O1 interface defines a set of roles that control access for all O-RAN deployments. Using these roles, an operator can provide segregation of duties, especially separating the management of the O-RAN NFs from user management.

- **O1_nacm_management:** Allows changes to the /nacm objects which includes the NACM Global Enforcement Controls.
- **O1_user_management:** Allows assignment and deletion of users and assignment of users to roles on the O1 node.
  - o Mandatory if the network device supports a local user store.
  - o Not provided if the network device does not support a local user store and requires all user/role information to be provided by an external authentication/authorization service.
- **O1_network_management:** Allows read, write, and execute operations on the <running> datastore and read, write, execute, and commit operations on the <candidate> datastore if <candidate> is supported. All operations on the /nacm objects are prohibited.
- **O1_network_monitoring:** Allows read operations on configuration data in the <running> datastore, except for the /nacm objects.
- **O1_software_management:** Allows installation of new software including new software versions.

8.4.1.3 O2

The O2 interface (O-RAN ALLIANCE O-RAN.WG6.O2-GAnP-v01.02 2022) provides APIs that are used by the SMO to manage the O-Cloud and to retrieve information about the O-Cloud infrastructure. The O2 interface follows the ETSI RESTful NFV MANO APIs (ETSI GS NFV-SOL 013 v3.3.1 2000) section 8 for interface security and requires OAuth 2.0 using TLS with Public Key Infrastructure X.509 (PKIX) mutual authentication to obtain an access token from the OAuth Authorization Server. As part of setting up the TLS tunnel for the access token request, the client and authorization server perform mutual authentication based on X.509 certificates and the client presents its client identifier. The TLS tunnel between the SMO and authorization server provides secure transmission of the access token. Subsequent requests to the O-Cloud APIs are also secured using TLS 1.2 or 1.3, thereby providing confidentiality, integrity, and replay protection. It should be noted that authentication with client secret (password) may be used to support older implementations of the RESTful NFV MANO APIs (prior to version 3.4.1).

8.4.1.4 E2

The E2 interface is currently the only O-RAN interface that uses IPSec to provide confidentiality, integrity, replay protection, and data origin authentication. Authentication with X.509 certificates is mandatory. Vendor products may also support authentication with preshared keys, but their use in production environments is strongly discouraged. The E2 interface uses the Stream Control Transmission Protocol (SCTP) with ASN.1 payload.

8.4.1.5 Open Fronthaul

The Open Fronthaul interfaces provide a NETCONF management interface (the Open Fronthaul M-plane interface) between the SMO and O-RU and the O-DU and O-RU plus control, user and signaling interfaces between the O-DU and O-RU (Open Fronthaul C-plane, U-plane, and S-plane).

8.4.1.5.1 Open Fronthaul M-Plane The Open Fronthaul M-plane (O-RAN ALLIANCE O-RAN-WG4.MP.0-v09.00 2022) interface is built on NETCONF and provides very similar capabilities to the O1 interface. Like the O1 interface, authenticity, integrity, and confidentiality can be enforced using TLS 1.2 or 1.3 with PKIX. They can also be enforced using SSHv2 with password-based authentication. The Open Fronthaul M-plane interface also enforces RBAC using NACM, though the roles are slightly different from the O1 roles.

- **sudo:** Has full read, write, and execute privileges for all O-RU modules.
- **smo:** Full service management and orchestration capabilities.
- **hybrid-odu:** Similar service management and orchestration capabilities to the smo role, enabling an O-DU to manage an O-RU in conjunction with an SMO.
- **nms:** Allows a dedicated external network management system to manage the O-RU.
- **fm-pm:** Allows the collection of fault and performance management data.
- **swm:** Allows installation of new software including new software versions.

8.4.1.5.2 Open Fronthaul C/U/S-Plane At the time of publication, message-level security controls on the C, U, and S planes of the Open Fronthaul interface are not provided. The confidentiality of messages on the U-plane is provided by the Packet Data Convergence Protocol (PDCP) defined in 3GPP. The data on the C and S planes is not considered sensitive, and there are considerable technical constraints that preclude message-level security at this time. The O-RAN Alliance continues evaluating message-level security technologies and will introduce them when they become feasible (O-RAN ALLIANCE O-RAN.WG4.CUS.0-v07.02 O-RAN 2022). Currently, the Open Fronthaul C, U, and S plane interfaces rely on securing the point-to-point, or physical or L1, LAN segments between physical O-RAN Open Fronthaul elements using the IEEE 802.1X-2020 protocol (IEEE 802.1X 2020). The O-RAN Open Fronthaul elements include not only the O-DU and O-RU (see Figure 8.1) but also the Xhaul transport network elements (TNEs) that share a point-to-point LAN segment with Open Fronthaul network elements, such as switches, fronthaul multiplexor (FHM), fronthaul gateway (FHGW), TNE, Primary Reference Time Clock and Telecom

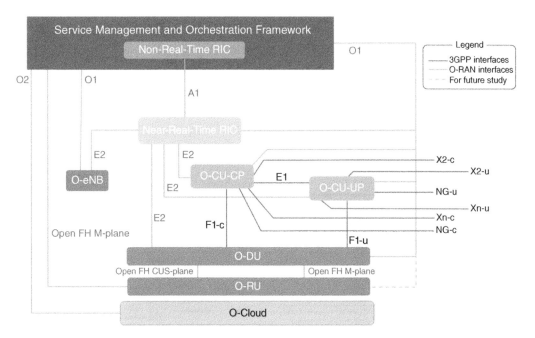

Figure 8.1 O-RAN high-level architecture.

Grand Master (PRTC/T-GM). The IEEE 802.1X-2020 protocol enables the Open Fronthaul elements a way to authenticate endpoints on a point-to-point LAN segment, detecting and reporting on new or broken segments, and blocking unused Ethernet ports in Open Fronthaul network elements.

Port-based network access control in the O-RAN Alliance Open Fronthaul comprises supplicant, authenticator, and authentication server entities described in (IEEE 802.1X). An authenticator is an entity that facilitates the authentication of other entities attached to the same LAN. A supplicant is an entity at one end of a point-to-point LAN segment that seeks to be authenticated by an authenticator attached to the other end of that link. Finally, an authentication server provides an authentication service to an authenticator that determines, from the credentials provided by the supplicant, whether the supplicant is authorized to access the services provided by the system in which the authenticator resides.

Every element on an O-RAN Open Fronthaul network supports both the supplicant functionality and the authenticator functionality. Port-based network access control between a supplicant and authenticator in an Open Fronthaul network uses the Extensible Authentication Protocol (EAP TLS) authentication as defined in (IEEE 802.1X) using X.509 certificates. The interface between an authenticator and the authentication server uses the RADIUS protocol (IETF RFC 2865 2000), and optionally may use the DIAMETER (IETF RFC 4072 2004) protocol. Only those Open Fronthaul network elements acting as a supplicant that have mutually authenticated with an authenticator are authorized to participate in the Open Fronthaul network.

8.4.1.6   R1

The R1 interface is a recently added O-RAN interface that formalizes communication between the non-RT RIC and the SMO (see Figure 8.2: SMO and non-RT RIC Architecture). The O-RAN Alliance plans to secure this interface using TLS 1.2 and 1.3 with mutual authentication using PKIX.

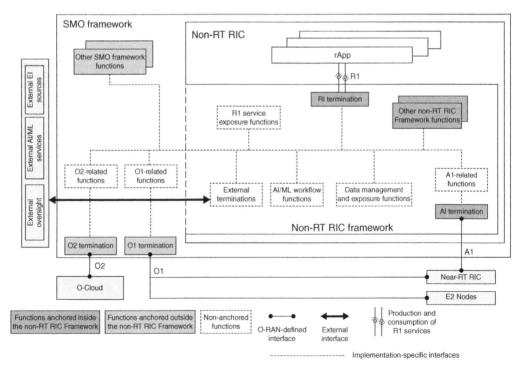

Figure 8.2   SMO and non-real-time RIC architecture.

8.4.1.7   3GPP Interfaces

3GPP interfaces supported by O-RAN, specifically the E1, F1-c, F1-u, NG-u, NG-c, X2-u, X2-c, Xn-c, Xn-u, use the 3GPP-mandated security controls to enforce authenticity, integrity, confidentiality, and least privilege.

8.4.2   Securing the O-Cloud

O-RAN deployments will be deployed on cloud infrastructures and leverage other virtualization techniques such as containerization with Kubernetes orchestration. The cloud is designed to support multitenancy, which is the use of a shared physical infrastructure by multiple customers. Following the recommendations of the ESF PART III, the O-RAN Alliance has identified the following mandatory controls that must be supported by any O-Cloud provider in order to ensure that workloads remain isolated from one another.

- Secure access controls for administrators.
- Secure API interface for tenants.
- Support for the confidentiality and integrity of data, including related metadata, at-rest, and in-transit.
- Strong tenant isolation – guaranteed confidentiality and integrity of processes and information sharing with only authorized parties (e.g. tenants).
- Support for the confidentiality and integrity of workload resource utilization (RAM, CPU, Storage, Network I/O, cache, and hardware offload) and the restriction of information sharing with only authorized parties.
- Prevent memory inspection by any actor other than the authorized actors for the entity to which memory is assigned (e.g. tenants owning the workload), for lawful inspection, and by secure monitoring services.
- Prevent the monitoring system from affecting the data confidentiality of the infrastructure, workloads, or user data.

The use of the cloud introduces the shared responsibilities model in which the cloud provider and the MNO are different entities responsible for different aspects of security. The cloud provider ensures that the physical and virtualization technologies are secured, while the MNO is responsible for the security of their tenants and their data.

8.4.3   Container Security

Container images are quickly becoming the default packaging mechanism for software components, as well as the preferred isolation strategy. MNOs should expect to see O-RAN components delivered as containers that are orchestrated by Kubernetes. The O-RAN Alliance will provide recommendations for vendors, MNOs, and administrators to follow when developing and deploying containerized O-RAN components. The primary aim is to isolate O-RAN containers, maintaining the integrity of each O-RAN component. At the time of publication, the O-RAN Alliance is investigating the CIS Benchmark recommendations for Docker (CIS 2021) and Kubernetes (CIS 2020) security.

8.4.4   O-RAN Software Security

Disaggregation means that O-RAN components, including x/rApps, will come from many providers: traditional telecom vendors, software vendors with limited telecom experience, data analytics companies, and open-source communities. Consistent software security practices for all these providers are imperative to minimize vulnerabilities and provide reliable vulnerability patching. At the time of publication, O-RAN requires all providers to digitally sign their software packages and that MNOs have the ability to verify the signature before deploying a component into the network. Signing allows an MNO to detect software tampering in their environment but does not ensure that the software is secure. At the time of publication, the O-RAN Alliance is developing a set of requirements for O-RAN providers using the NIST Secure Software Development Framework as a guide (Souppaya et al. 2022). Specific areas of focus will be designing in securable software, protecting software from tampering at all stages of development, testing for known and common vulnerabilities, testing all code branches (code coverage), staying current with all third-party software in their code, developing regular (and frequent) patching cycles, and responding to new and zero-day vulnerabilities in their code.

8.4.5   Software Bill of Materials (SBOM)

A software bill of materials (SBOM) is a list of all third-party software used in an application. It is recognized as being a fundamental artifact produced by a mature software development life cycle. An SBOM is invaluable in detecting if an application has exposure to a vulnerability. The US Department of Commerce (DoC) and the National Telecommunications and Information Administration (NTIA) define SBOM as "a formal record containing the details and supply chain relationships of various components used in building software." O-RAN strongly recommends that providers include an SBOM with all O-RAN components and that the SBOM be compliant with the NTIA recommendations (US DoC and NTIA 2021).

SBOM data fields
- The SBOM must contain as minimum fields: Supplier Name, Component Name, Version of the Component, Other Unique Identifiers, Dependency Relationship, Author of the SBOM Data, and Timestamp.

SBOM automation
- The interoperable data formats used to generate and consume SBOMs must be either Software Package Data eXchange (SPDX) (https://spdx.dev), CycloneDX (https://cyclonedx.org), or Software Identification (SWID) (Waltermire 2016) tags.

SBOM practices
- The cryptographic hash and digital signature of the SBOM may be defined in the contractual agreement between the software supplier and customer.
- The terms for access control to the SBOM may be defined in the contractual agreement between the software supplier and customer.
- Depth is the level of upstream suppliers maintained in the SBOM. An SBOM should contain all primary (top-level) components, with all their transitive dependencies listed. At a minimum, all top-level dependencies must be listed with enough detail to seek out the transitive dependencies recursively.
- The commercial software supplier is not obligated to make the SBOM publicly available.

8.5   Recommendations for Vendors and MNOs

O-RAN security depends on both O-RAN component vendors and MNOs. The vendors have to implement all mandatory (and ideally optional) security requirements, use secure software development life cycles, perform rigorous security testing of their products, integrate with standard security protocols, and rapidly respond to vulnerabilities in their products. MNOs have to use all security features, update regularly, and monitor their O-RAN deployments for changes to security configurations.

This list of vendor practices should not be considered exhaustive, but a starting point.

- Adopt secure development practices
  - Use continuous integration and continuous development (CI/CD) with continuous regression testing and software security auditing SHOULD be implemented
  - Scan all source code as it is developed (SAST, DAST, and SCA)
  - Create automated test suites that cover all code branches
  - Test for security vulnerabilities using techniques such as fuzzing, large inputs, code evaluation
  - Ensure that all software and hardware produced meets the 3GPP Security Assurance Specifications (SCAS), Cyber Act, General Data Protection Regulation (GDPR)
  - Use the latest General Availability (GA)-supported versions of all third-party software
  - Remove all unsupported third-party software
  - Use only industry-recommended cryptography and security protocols
  - Understand the provenance of all included code by using a software bill of materials (SBOM)
- Create a response process for bugs and vulnerabilities
  - Receive advisories from major information security organizations about vulnerabilities in third-party products
  - Inform major information security organizations about vulnerabilities in their products with timelines to remediate
  - Inform customers promptly of newly found bugs and vulnerabilities
  - Provide customers recommendations for compensating controls while remediating the vulnerabilities and bugs
  - Provide customers with a timeline for delivering fixes
  - Prioritize vulnerability and bug fixes over new features
  - Deliver fixes efficiently
  - Make vulnerability response metrics available to customers
- Secure the supply chain
  - Provide an up-to-date and digitally signed SBOM for each product
  - Require an SBOM for all third-party code used in software products
  - Require a bill of materials (software and hardware) for all third-party components in your products

Similarly, the list of MNO practices is a starting point for operating a secure O-RAN deployment

- Establish clear security requirements for vendors
  - Require Service Leval Agreements (SLAs) for vulnerabilities and bug fixes from all vendors
  - Require signed SBOMs for all software products and signed SBOMs and hardware BOM for all hardware-based products
  - Review vendor vulnerability response metrics
- Inventory control
  - Maintain digital, searchable SBOMs and hardware BOMs for all products in the application and system inventory
- Patch management
  - Develop a regular and frequent patch management schedule (NIST SP 800-40 2013)
  - Deploy compensating controls for known vulnerabilities
  - Deploy vulnerability fixes as rapidly as possible

- Security controls
  - Enable all mandatory and optional security controls
  - Use only industry-recommended cryptography and security protocols
  - Use a secure PKI infrastructure
  - Harden all compute, storage, network, and virtualization platforms
  - Harden all virtual and physical machines by disabling all unused ports and services
  - Use a hardened Identity and Access Management (IAM) solution with mature identity life cycle management to enforce authentication and access control on interfaces
  - Use an enterprise-grade log management system to collect and monitor security events
    - Log all relevant events with sufficient metadata to enable analysis
    - Perform continuous log analysis to identify anomalous behavior
  - Implement continuous monitoring of all information resources to verify the expected security state is maintained throughout the lifecycle of deployed O-RAN components
  - Secure software repositories to prevent software tampering
- Security practices
  - Perform regular penetration tests
  - Use zero touch deployment and management to reduce ad hoc management
  - Use change control process for all modifications to the environment
- Security virtualization strategy
  - Ensure robust isolation of each O-RAN component

8.6 Conclusion

O-RAN security is evolving with the platform. The introduction of new, well-defined open interfaces not only increases the defined threat surface but also provides the opportunity to define and implement standard and modern security controls. With or without O-RAN, AI/ML will be used to operate the RAN. The use of these techniques via the non- and near-RT RIC open standards, enables vendors and operators to control AI/ML security threats instead of relying on vendors to secure proprietary solutions. Ultimately, the more open architecture will yield a more secure RAN that can rapidly incorporate new security practices.

Bibliography

3GPP 33.210 (2022). *3GPP TS 33.210: Technical Specification Group Services and System Aspects; Network Domain Security (NDS); IP Network Layer Security.*

CIS (2020). *CIS Kubernetes Benchmark, v1.6.0.*

CIS (2021). *CIS Docker Benchmark, v1.3.1.*

CISA, NSA (2021). *Security Guidance for 5G Cloud Infrastructures, Part II: Securely Isolate Network Resources.* https://www.cisa.gov/sites/default/files/publications/Security_Guidance_For_5G_Cloud_Infrastructures_Part_II_Updated_508_Compliant.pdf.

CISA, NSA (2021). *Security Guidance for 5G Cloud Infrastructures, Part III: Data Protection.* https://www.cisa.gov/sites/default/files/publications/Security_Guidance_For_5G_Cloud_Infrastructures_Part_III_508_Compliant.pdf.

CISA, NSA, Office of the Director of National Security (2021). *Potential Threat Vectors to 5G Infrastructure*. https://media.defense.gov/2021/May/10/2002637751/-1/-1/0/POTENTIAL%20 THREAT%20VECTORS%20TO%205G%20INFRASTRUCTURE.PDF.

ETSI GS NFV-SOL 013 v3.3.1 (2000). *Protocols and Data Models; Specification of common aspects for RESTful NFV MANO APIs*.

IEEE 802.1X (2020). *IEEE Standard for Local and Metropolitan Area Networks--Port-Based Network Access Control*, in IEEE Std 802.1X-2020 (Revision of IEEE Std 802.1X-2010 Incorporating IEEE Std 802.1Xbx-2014 and IEEE Std 802.1Xck-2018), pp. 1–289, 28 Febuary 2020, doi: 10.1109/ IEEESTD.2020.9018454.

IETF RFC 2818. (2000). *HTTP over TLS*. https://datatracker.ietf.org/doc/html/rfc2818.

IETF RFC 2865 (2000). *IETF RFC 2865. Remote Authentication Dial In User Service (RADIUS)*. June 2000. https://www.rfc-editor.org/rfc/rfc2865.html.

IETF RFC 4072 (2004). *Diameter Extensible Authentication Protocol (EAP) Application*. https:// datatracker.ietf.org/doc/html/rfc4072.

IETF RFC 4252 (2006). *The Secure Shell (SSH) Authentication Protocol*. https://tools.ietf.org/html/ rfc4252.

IETF RFC 4303 (2005). *IP Encapsulating Security Payload (ESP)*. https://datatracker.ietf.org/doc/html/ rfc4303.

IETF RFC 5246 (2008). *The Transport Layer Security (TLS) Protocol Version 1.2*. https://datatracker.ietf. org/doc/html/rfc5246.

IETF RFC 6241 (2011). *Network Configuration Protocol (NETCONF)*. https://tools.ietf.org/html/ rfc6241.

IETF RFC 6749 (2012). *The OAuth 2.0 Authorization Framework*. https://tools.ietf.org/html/rfc6749.

IETF RFC 8341 (2018). *Network Configuration Access Control Model*. https://tools.ietf.org/html/ rfc8341.

IETF RFC 8446 (2018). *The Transport Layer Security (TLS) Protocol Version 1.3*. https://datatracker.ietf. org/doc/html/rfc8446.

National Security Agency (NSA) and Cybersecurity and Infrastructure Security Agency (CISA) (2021). *Security Guidance for 5G Cloud Infrastructures, Part I: Prevent and Detect Lateral Movement*. https:// media.defense.gov/2021/Oct/28/2002881720/-1/-1/0/SECURITY_GUIDANCE_FOR_5G_CLOUD_ INFRASTRUCTURES_PART_I_20211028.PDF.

O-RAN ALLIANCE O-RAN.SFG.O-RAN-Security-Requirements-Specifications-v03.00 (2022). *O-RAN Security Requirements Specification 3.0*. https://www.o-ran.org/specifications.

O-RAN ALLIANCE O-RAN.SFG.O-RAN-Threat-Model-v03.00 (2022). *O-RAN Security Threat Modeling and Remediation Analysis 3.0*. https://www.o-ran.org/specifications.

O-RAN ALLIANCE O-RAN.SFG.Security-Protocols-Specifications-v03.00 (2022). *Security Protocols Specifications 3.0*. https://www.o-ran.org/specifications.

O-RAN ALLIANCE O-RAN.WG10.O1-Interface.0-v07.00 (2022). *O-RAN Operations and Maintenance Interface Specification 7.0*. https://www.o-ran.org/specifications.

O-RAN ALLIANCE O-RAN.WG4.CUS.0-v07.02 O-RAN (2022). *O-RAN Control, User and Synchronization Plane Specification 7.0*. https://www.o-ran.org/specifications.

O-RAN ALLIANCE O-RAN.WG6.O2-GAnP-v01.02 (2022). *O-RAN O2 Interface General Aspects and Principles 1.02*. https://www.o-ran.org/specifications.

O-RAN ALLIANCE O-RAN-WG4.MP.0-v09.00 (2022). *O-RAN Management Plane Specification 9.0*. https://www.o-ran.org/specifications.

Rose, S., Borchert, O., Mitchell, S., and Connelly, S. (2020). *NIST Special Publication 800–207: Zero Trust Architeecture*. https://nvlpubs.nist.gov/nistpubs/SpecialPublications/NIST.SP.800-207.pdf.

Souppaya, M., and Scarfone, K. (2013). *NIST Special Publication 800–40: Guide to Enterprise Patch Management Planning Revision 3: Preventive Maintenance for Technology*. https://nvlpubs.nist.gov/nistpubs/SpecialPublications/NIST.SP.800-40r3.pdf.

Souppaya, M., Scarfone, K., and Dodson, D. (2022). *Secure Software Development Framework (SSDF) Version 1.1: Recommendations for Mitigating the Risk of Software*.

U.S. DoC and NTIA (2021). *The Minimum Elements for a Software Bill of Materials (SBOM), Pursuant to Executive Order 14028 on Improving the Nation's Cybersecurity*.

Waltermire, D. (2016). *Guidelines for the Creation of Interoperable Software Identification (SWID) Tags*. http://dx.doi.org/10.6028/NIST.IR.8060.

9

Open RAN Software

David Kinsey[1], Padma Sudarsan[2], and Rittwik Jana[3]

[1] AT&T, Seattle, WA, USA
[2] VMWare, Palo Alto, CA, USA
[3] Google, New York City, NY, USA

9.1 Introduction

In the past, software in the telecommunications sector has been largely proprietary and developed for specific hardware due to stringent performance requirements. OpenStack is a free, open standard cloud computing platform. It is mostly deployed as Infrastructure-as-a-Service (IaaS) in both public and private clouds where virtual servers and other resources are made available to users. The development of OpenStack has heavily used network virtualization in the mobile packet core based on commodity hardware. It has also moved on to support not just the decoupling of the Control and User Plane aspects of the functions but also their coupling to a specific hardware implementation. The same challenges exist for the development of Open RAN and virtualized RAN software infrastructures. With the introduction of containerization and solutions such as Kubernetes and the inclusion of accelerators into the cloud fabric, the performance requirements of being able to handle real-time RAN workloads can now be met through a cloud-native approach.

The microservice architecture of cloud-native applications promotes the reuse of common functions. This has been accelerated by open-source developments that can solve the nondomain-specific aspects of a network function (NF). As a result, Open RAN technology providers are increasingly using open-source software under the hood. Open-source platform components (like Prometheus, fluentd, etc.) can/will also play a crucial role in the orchestration and management of cloud deployments.

This chapter discusses the various open-source projects that are focused on RAN. Specifically, we discuss O-RAN software community (OSC) and the various projects and use cases and APIs that are in progress. We also describe some Open Network Automation Platform (ONAP) use cases that are aligned with O-RAN and the efforts needed to harmonize among these two communities. ONAP is an open-source, orchestration, and automation framework. Finally, we touch on other open-source communities (e.g. Anuket, OAI, ONF, PAWR, POWDER, etc.) that are gaining momentum and are helping to create a rich ecosystem of intelligent RAN applications.

Open RAN: The Definitive Guide, First Edition. Edited by Ian C. Wong, Aditya Chopra, Sridhar Rajagopal, and Rittwik Jana.
© 2024 The Institute of Electrical and Electronics Engineers, Inc. Published 2024 by John Wiley & Sons, Inc.

9.2 O-RAN Software Community (OSC)

The OSC (https://wiki.o-ran-sc.org/) is a collaboration between the O-RAN ALLIANCE (O-RAN) and Linux Foundation (LF) with the mission to support the creation of software for the radio access network (RAN). Some of the contributions to the OSC contain intellectual property (IP) or patents from various O-RAN members. This software is made available via the O-RAN Software License. Other contributions are made available through the Apache 2.0 License.

9.2.1 OSC Projects

The OSC is aligned to deliver the NFs in order to enable a complete RAN stack. It is therefore loosely guided by the O-RAN Architecture Description. Figure 9.1 represents the O-RAN logical architecture.

9.2.2 The Service Management and Orchestration (SMO) Framework

The SMO is decomposed into three different projects:

OAM: The OAM project (Operations and Maintenance – OAM – Confluence [https://wiki.o-ran-sc.org/pages/viewpage.action?pageId=3605135]) is mostly code/images reused from the ONAP platform described later. It provides the base connectivity between the SMO and the network over the A1 and O1 interfaces. This primarily consists of the Netconf client (SDN-R) and the FCAPs collector (VNF Event Streaming (VES)) with their specific OSC configurations. At the time of this writing, the O2 is just emerging and it is unclear if there will be adapters required by the OAM or if the SMO will directly integrate with the infrastructure (INF) project described later.

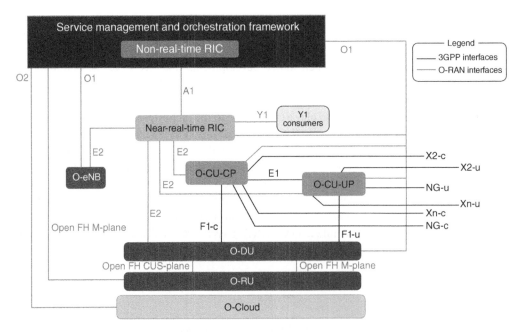

Figure 9.1 O-RAN logical architecture *Source:* O-RAN.WG1.OAD-R003-v08.00, Nov 2022

Non-RT RIC: The non-real-time RIC (RAN Intelligent Controller) is an Orchestration and Automation function described by the O-RAN Alliance for non-RT intelligent management of RAN functions. The primary goal of the non-RT RIC is to support non-RT radio resource management, higher layer procedure optimization, policy optimization in RAN, and providing guidance, parameters, policies, and AI/ML models to support the operation of near-real-time (RT) RIC functions in the RAN to achieve higher-level non-RT objectives. Non-RT-RIC functions include service and policy management, RAN analytics, and model training for the near-RT RICs. The non-RT RIC project provides concepts, specifications, architecture, and reference implementations as defined and described by the O-RAN Alliance architecture.

    The non-RT RAN Intelligent Controller (NONRTRIC) (Non-RealTime RIC [NONRTRIC] – Non-Realtime RIC – Confluence [https://wiki.o-ran-sc.org/display/RICNR]) is intended to provide an extensible framework for microservices (rAPPs) to be used to implement closed loop controls for the RAN with latencies greater than one second. Where the non-RT RIC is integral to the SMO, the rAPPs are expected to be provided by other vendors who specialize in specific RAN optimizations or controls. Therefore, there is a standard interface, the R1, being specified by O-RAN and implemented in the NONRTRIC project. The project also includes some sample rAPPs which are intended to demonstrate how to develop an rAPP or demonstrate capabilities requested by O-RAN.

SMO: The SMO project comprises the higher-level functions of the SMO framework. Although OSC leverages some ONAP components, the SMO is not solely an ONAP implementation. The SMO adopts the O1 VES formats allowing a single collector provided and configured by the OAM project to receive many of the O1 telemetry messages. The SMO uses several other open-source tools which are easily integrated (e.g. Kafka for a Message Bus, with an ONAP DMaaP adapter). It also supports an InfluxDB for time series persistence of telemetry and LogStash for analysis of telemetry data. Future work is intended to incorporate TACKER from OpenStack for some of the NF life cycle orchestration, cooperating with HELM and Kubernetes.

9.2.3 Near-RT RIC (RIC)

The near-RT RIC is a near-real-time RAN controller deployed in a cloud infrastructure at the edge of the RAN, which is responsible for fine-grained RRM decisions of control-plane and user-plane procedures (mostly UE-specific) pertaining to functions across the RAN protocol stack near-RT granularities with low-latency closed-loop control loops ranging from 10 ms to 1 s over the O-RAN-standardized E2 interface (See Chapter 3).

    The near-RT RIC is decomposed into two different projects:

RIC: The RIC project (Near Realtime RAN Intelligent Controller [RIC] – RIC Platform – Confluence [https://wiki.o-ran-sc.org/pages/viewpage.action?pageId=1179659]) is the largest project in the OSC. As a new element in the RAN, it provides the newest functionality. The O-RAN RIC is composed of a framework, this project, and extensible third party applications (xAPPs). The near-RT RIC is the managed element in the network. Applications are deployed to integrate with the platform and become managed through the O1 interface to the near-RT RIC. The near-RT RIC also has the A1 interface to the SMO, and the E2 interface to other RAN functions. These other functions are namely the O-CU-UP, O-CU-CP, and O-DU but are often referred to as E2 Nodes.

RICAPP: The RICAPP (RIC Applications [RICAPP] – RIC Applications – Confluence [https://wiki.o-ran-sc.org/pages/viewpage.action?pageId=1179662]) project provides examples of xAPP implementations. There are a variety of application behavior types ranging from data collection to AI/ML-enabled RAN control functions. Each xAPP in the project has its own repository (repo). Some of the xAPPs work together as a set to enable a specific optimization.

9.2.4 O-CU-CP and O-CU-UP

The OCU project (O-RAN Central Unit [OCU] – O-RAN Central Unit – Confluence [https://wiki.o-ran-sc.org/pages/viewpage.action?pageId=1179674]) was initially started but stopped due to a lack of contributor resources. Today end-to-end testing is performed by using a binary image test stub provided by Radisys.

9.2.5 O-DU Project

The O-DU has stringent timing interactions dependent on the spectrum band used by the radio. Only the low bands of the sub-6 spectrum frequencies are expected to be supportable on standard hardware platforms. Higher frequencies will require specific accelerators to be incorporated into the edge cloud (O-Cloud) in order to support a software defined O-DU. In OSC, the O-DU is developed as a network appliance which registers itself with the SMO as a physical NF. It is decomposed into two different projects representing the higher level functions and the lower level functions communicating through the FAPI interface defined by the small cell forum (SCF) (5G FAPI Suite – Small Cell Forum):

ODUHIGH: The ODUHIGH project (O-DU High Overview – O-RAN Distributed Unit – Confluence [https://wiki.o-ran-sc.org/pages/viewpage.action?pageId=1179671]) implements the functional blocks of L2 layers of a 5G NR protocol stack in Stand Alone (SA) mode. The L2 layers are NR RLC (Radio Link Control), NR MAC (Medium Access Control) and NR Scheduler. These layers aid in the transmission of traffic from the UE to the 5G core network (CN). The 5G NR RLC layer provides services for transferring the control and data messages between MAC layer and O-CU (via DU App). The 5G NR MAC layer uses the services of the NR physical layer (O-DU Low) to send and receive data on the various logical channels using multiplexing and demultiplexing techniques. The 5G NR SCH layer allocates resources in UL and DL for cell and UE-based procedures.

ODULOW: The ODULOW project (O-DU Low – O-RAN Distributed Unit – Confluence [https://wiki.o-ran-sc.org/display/ORANDU/O-DU+Low]) in OSC brings in the binary from the Open-Source Flexran (GitHub – intel/FlexRAN) by Intel. FlexRAN is Intel 4G and 5G baseband PHY Reference Design, which uses Xeon® series Processor with Intel Architecture. It is the Flexran dependency on specific hardware that drives the Appliance Model of the O-DU.

9.2.6 O-RU

The O-RU is always expected to be a physical NF and therefore a software project was never started. OSC uses a UE simulator and O-RU emulator. These capabilities are provided by test equipment from various vendors.

9.2.7 O-Cloud

The INF (infrastructure) project [https://wiki.o-ran-sc.org/pages/viewpage.action?pageId=1179726] provides open-source reference implementation of Edge Cloud infrastructure according to the O-RAN WG6 specification to be used with the other O-RAN OSC projects such as O-CU, O-DU and near-RT RIC. The O-Cloud is a distributed cloud with resources in multiple locations. Some far-edge locations might be limited in capacity and therefore a SIMPLEX model is supported where the cloud management, control, and deployment planes are implemented in a single server. Other far edge locations may be able to support a more robust implementation which includes some failover capabilities where two servers (DUPLEX) are used. The general edge cloud cluster is typically more than two servers. In this case, the DUPLEX+ model is used where two servers are dedicated to the management and control plane and other dedicated worker nodes are added to the deployment plane.

9.2.8 The AI/ML Framework

The main goal of the AI/ML Framework (AIMLFW) project [https://wiki.o-ran-sc.org/pages/viewpage.action?pageId=57377181] is to create an end-to-end framework, which manages and performs AI/ML-related tasks including training, model deployment, inference, and model management. This project was motivated by the study presented in the technical report by O-RAN alliance Working Group 2 (WG2) dealing with non-RT (NRT) RIC.

9.2.9 Support Projects

There are several supporting projects in the OSC.

SIM: The SIM project (SIM – Confluence [https://wiki.o-ran-sc.org/pages/viewpage.action?pageId=9568290]) implements either end of an interface thus allowing individual testing and in some cases, end-to-end testing of O-RAN capabilities.

DOC: The DOC project (Documentation – Confluence [https://docs.o-ran-sc.org/en/latest/]) provides the release documentation for all projects. It contains the how-to documentation if you are interested in using or executing OSC software.

INT: The INT (integration) project (Integration and Testing – Confluence [https://wiki.o-ran-sc.org/pages/viewpage.action?pageId=1179695]) performs the release integration testing for all projects. It also contains the Open Test Framework which allows the projects to define repeatable tests for implementation verification or end-to-end demonstrations. It also allows for Test to be defined for Specification Conformance or Interoperability to be defined as they are specified by O-RAN.

9.3 Open Network Automation Platform (ONAP)

Open Network Automation Platform (ONAP) is an open-source, orchestration, and automation framework. The goal of the project is to develop a widely used platform for orchestrating and automating physical and virtual network elements, with full life cycle management. As of January 2018, ONAP became a project within the LF Networking Fund, which consolidated membership across multiple projects into a common governance structure. ONAP was not developed specifically for O-RAN. However, since the formation of O-RAN, it has had an alignment program to enhance the platform with features critical for a successful SMO.

9.3.1 Netconf/YANG Support

Improvements to the Controller framework (SDN-R) were done to map Netconf notifications from the FH (Fronthaul) M-Plane interface to forward fault and performance from O-RU devices being managed in a Hybrid mode, as specified in the OAM Architecture Specification allow ONAP to not only support O1 but FH M-Plane elements as well. Additional modifications were added to allow the NETCONF session to not be continuous as this might become a cost burden in extremely large networks where the number of O-RU can be in the millions due to the extremely small footprint size of millimeter wave (mmW) cells, for example.

9.3.2 Configuration Persistence

ONAP did not store the current or past configuration of the elements in a repository. This would then require all control functions needing the current device state to have to query the device. Due to low latency control loops, this would become a performance burden on the cell itself. A project was developed to maintain a synchronized copy of the element configuration in ONAP. Therefore, all optimizers can utilize that information as part of its analysis. This entails subscribing to configuration change events, as elements such as the O-RU might be changed by the O-DU, the SMO, and in the future, possibly another O-DU which is sharing the O-RU.

Another critical aspect of configuration is the relationship between Managed Element and Managed Function. The O1 interface terminates at the Managed Element. The optimizers tune the Managed Functions. In O-RAN, there is no constraint to the number or type of Managed Functions contained in a Managed Element. Therefore, the composition of the Managed Element, discovered through the configuration data, is critical for all O-RAN management functions.

9.3.3 VES Support

VES was a concern to O-RAN. It lacked the ability to allow native SDO defined data, such as 3GPP, be passed. Since that data changes with 3GPP releases, a way was needed to allow different schemas as not all elements are expected to be updated to the same release at a given point in time. A new domain for standard defined schemas, identified by their namespace value and schema URL, was introduced. This allows the VES listener to still validate an event to a specific schema prior to indicating acceptance to the producing network element. Not only does this future prove the VES capabilities, but it also allows ONAP to have a geo-distributed, load-balanced pool of VES collectors providing high-availability and high-transaction volume rates to the network data producers.

9.3.4 A1 Support

Some OSC prototypes for A1 policy support were hardened and incorporated into the ONAP controller framework. The set of procedures associated with the A1 interface constitute the A1 Application Protocol (A1AP), which enables a direct association between the non-RT RIC and the near-RT RIC over A1. The A1 termination function routes rApp-related messages to the target rApp for generating services like A1-P, A1-EI, etc. This enabled ONAP to also support the evolving A1 interface. This activity is still ongoing as the initial A1 for policy is expanded to support Enrichment Information as well. Future work should be expected when the specification for ML updates over A1 is specified by O-RAN in A1-ML.

9.3.5 Optimization Support

As the basic FCAPS was incorporated into ONAP and Slice Management began to emerge in OSC, concerns for Optimization loops emerged. Specifically, the fact that the O1 interface only allowed batch (file-based) transmission or Streaming interfaces over Websockets encoded by ASN.1 to be transmitted. File based would not support low-latency control loops and issues were discovered with the streaming data as specified by 3GPP and adopted by O-RAN.

A specific issue was the coupling of the PM Job creation with the stream content. Additionally, the receiver needed to know exactly when the job, and thus the stream, was modified. Therefore, there exists a window in which data corruption can occur in the stream data. To correct this, work is being done to identify the errors and correct them in O-RAN and hopefully harmonize that with 3GPP.

Several changes are in discussion in the O-RAN community:

- The first change addresses the ASN.1 streaming performance requirements in the O1 implementation. A change has been discussed and accepted to the O-RAN O1. However, although 3GPP has specified that Google Protobuf (GPB) could also be used, it has not defined the schema. It is therefore, currently still impossible to create a 3GPP compliant GPB implementation.
- Today 3GPP requires the PM Stream consumer to receive just the measurements and map those to field names based on a separately managed list of field names using a common index value between the two independent lists. There is no coordination when a PM Job is modified and, therefore, even though the index has changed somewhere in the stream, the consumer is never notified of where the event happened. Some measurement names are extremely long. However, ONAP introduced the concept of a Performance Measure (PM) Dictionary. This provides an identity value for each measure. Therefore, passing the identity value paired with a measurement value is not an extreme burden for ASN.1 or GPB. The stream data, having both values and not relying on a common index value, removes the vulnerability.
- Decoupling of PM Job Creation with PM Job Subscriptions. Currently jobs are defined with the reporting option as part of its parameters. This coupling is not only unnecessary but it restricts the SMO as more dynamicity is required. 3GPP only allows a PM Metric to be contained in one job. However, if two different SMO analytics need the data at different availability, the SMO will have to modify existing PM jobs and downstream flows to meet the demands of the analytic consumers. This complexity can be avoided by simply decoupling the PM Job which makes the measurements available from the subscription job which is used to feed the data to the SMO according to its current requirements. Once decoupled, then a subscription could also subscribe to data from different PM Jobs and thus multiplex PM Measures into common subscriptions based on latency and data volume.
- Performance data is the cornerstone for network optimization. However, in the O1, the message-based reporting was not specified. The current discussions are to enable a message-based notification of PM data. This allows easy initial prototyping for low latency, and relatively low-volume data to be acquired. If the data sent in a message is commensurate with data in a stream, then migration to streams becomes simplified without the need for a major impact to the analytics. Additionally, it becomes possible to inject message data with streaming data into an analytic that might combine them, allowing for statistical metrics to be calculated potentially outside of an inference engine which compares streaming data to statistical data values.

9.4 Other Open-Source Communities

Nephio is a Kubernetes-based intent-driven automation of NFs and the underlying infrastructure that supports those functions. It allows users to express high-level intent, and provides intelligent, declarative automation that can set up the cloud and edge infrastructure, render initial configurations for the NFs, and then deliver those configurations to the right clusters to get the network up and running.

OpenAirInterface Software Alliance (OSA): The OSA is the home of OpenAirInterface, an open software that gathers a community of developers from around the world, who work together to build wireless cellular RAN and CN technologies.

The OSA Alliance is responsible for:

- the development roadmap,
- the quality control,
- the promotion of the OAI software packages, deployed by our academic and industrial community for varied use cases.

ANUKET: Anuket [https://wiki.anuket.io/], formed out of OPNFV and the Cloud iNfrastructure Telco Taskforce (CNTT), is the transformative organization needed to create a common understanding and new capability for infrastructures across the telecom industry and to plot our collective future. To further amplify its mission, Anuket works in partnership with GSMA and other standard bodies; open-source communities including CNCF, ONAP, and OpenStack; and industry-leading network operators and NFVI/VNF suppliers, in a truly global and collaborative effort.

CNCF: LF is the parent of CNCF. It is one of the LF's largest subfoundations. CNCF is the open-source, vendor-neutral hub of cloud-native computing, hosting projects like Kubernetes and Prometheus to make cloud-native universal and sustainable. The figure given here shows the various companies that make up the cloud-native landscape.

9.5 Conclusion

This chapter provided a glimpse of the various open-source projects related to RAN, cloud infrastructure, and orchestration and automation frameworks. A large part of the chapter discussed the projects in the OSC project within the LF. Other sister open-source projects under LF were briefly described (e.g. ONAP, Nephio, and CNCF). Finally, we touched on other open-source communities (e.g. Anuket, OAI, ONF, PAWR, POWDER, etc.) that are gaining momentum and are helping to create a rich ecosystem of intelligent RAN applications.

Bibliography

CNCF. The Open Source, Vendor-Neutral Hub of Cloud Native Computing. https://www.cncf.io.
Nephio. Intent-Driven Automation of Network Functions. https://wiki.nephio.org/display/HOME/Overview+of+Nephio.
ONAP. Open Networking and Automation Platform. https://www.onap.org.
The Open Air Interface Software Alliance. https://openairinterface.org.
The O-RAN Working Group 2 AI/ML Workflow Description and Requirements. O-RAN.WG2.AIML-v01.03 https://oranalliance.atlassian.net/wiki/download/attachments/120521394/O-RAN.WG2.AIML-v01.03.pdf, Oct. 2021.
The O-RAN Architecture Description Document. O-RAN.WG1.O-RAN-Architecture-Description-v08.00, Nov. 2022.
The O-RAN Software Community. https://wiki.o-ran-sc.org/display/ORAN/O-RAN+Software+Community.
The O-RAN Software Community Development Repository. https://oran-osc.github.io.

10

Open RAN Deployments

Sidd Chenumolu

DISH Network, Englewood, CO, USA

10.1 Introduction

This chapter aims to provide both an overview and detailed considerations to deploy a disaggregated, virtualized network based on O-RAN principles. In addition to O-RAN ALLIANCE (O-RAN) and 3GPP standards, details on virtualization and cloud-native implementation of RAN NFs are described to enable network operators to design and build a resilient and reliable O-RAN based network. O-RAN does not mandate any specific implementation of the network components, given the current state and the direction of the industry, it is fair to assume that RAN functionality will be implemented using commercial off-the-shelf (COTS) infrastructure instead of purpose-built application-specific integrated circuit (ASIC) hardware. Although the virtualized RAN applications can be both a virtual machine (VM)-based or container-based solution – this chapter uses the Kubernetes-based containerized deployment as an example. The principles and information provided on Open RAN deployment are agnostic of the frequency and duplexing schemes, and can be applied to both FDD and TDD as well as FR1 and FR2 as described in 3GPP (3GPP TS 38.101-1 2022).

O-RAN-based networks provide maximum flexibility and choice to operators to select not only the best network vendor products and solutions but also the best-of-breed components that power up the RAN network. For the first time, with Open RAN, mobile network operators (MNOs) are able to design, architect, and acquire solutions according to their needs instead of accepting a solution that is designed for the mass market and then designing their network around it. The RAN is no longer a black box solution – the Open RAN approach to disaggregation enables MNOs to get deep visibility and have full control of the composition of the RAN and its components to make informed buying decisions instead of being handed down a solution. Combined with the power of virtualization and cloud, Open RAN allows operators to have different solutions for small, micro, and macro deployments with different software, hardware, and feature sets and the ability to swap vendors if needed. There is no lock-in. One may argue that we lose efficiency, which is probably true in some cases, but flexibility and choice significantly outweigh the disadvantages of a vertically integrated solution. Contrary to popular belief or misconception, Open RAN is not just a lower-cost alternative solution to traditional RAN, it is about right-sizing, programmability, choice, and more importantly, openness. All this without giving up on performance or services. Open RAN can be deployed on both brownfield and greenfield networks, both as a replacement or a new radio access technology (RAT) addition.

This chapter is divided into three main sections – Network Architecture, Network Planning, and Network Deployment. The first section on Network Architecture describes the considerations for selection and designing Network Components such as radio unit (RU), distributed unit (DU), centralized unit (CU), and their management. The second section on Network Planning goes into details about RF Planning, Dimensioning, Timing design, and Cloud. The third section on Network Deployment describes different scenarios of Open RAN deployments and network management.

By the end of this chapter, it is expected that the reader will have developed a good understanding of the key network components, and design aspects that would enable the deployment of an O-RAN-based network.

10.2 Network Architecture

This section describes various network components and functions that are needed for O-RAN deployment. 5G is used as an example here. A high-level architecture diagram is shown in Figure 10.1 illustrating the placement of various 5G RAN and Core functions in a distributed manner. A cluster of RUs is connected to virtualized DUs (vDUs) which, in turn, are connected to virtualized Centralized Units (vCUs) placed at Edge Data Center (EDC). The vCU is further divided into separate instances of Control Plane and User Plane shown as vCU-CP and vCU-UP. The user plane aspect vCU-UP is connected to User Plane Function (UPF) and thereby to the internet. The control plane vCU-CP is connected to the 5G Core (5GC) and IP Multimedia Subsystems (IMS) at the Central Data Center to manage all control plane and multimedia communications services including PSAPs (Public Safety Answering Points), Law-Enforcement Agencies (LEAs), Roaming, and other public telephone networks (PSTNs).

The intent of this section is not to just describe the theoretical definitions of the network components but rather to highlight some of the considerations that are needed for deployment and operations. It is assumed that the reader has a good understanding of all O-RAN components already.

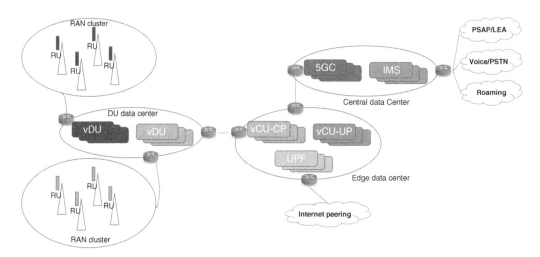

Figure 10.1 High-level E2E architecture.

If the NFs are virtualized, an operator may use a combination of private and public cloud options for its 5G deployment. This drawing depicts, at a high-level, the full network footprint that can even be implemented as a hybrid cloud strategy distributing functions across on-prem and even multiple cloud providers.

There are more technical details on each network component given as follows but an Operator should also overlay other business goals such as Sustainability and Carbon Neutrality to make selections.

10.2.1 Network Components

10.2.1.1 Antenna

Although the antenna is not an O-RAN component, it is a key factor to the network. For most deployments, except for massive MIMO (mMIMO), Antenna is a passive function that is pretty much agnostic of how the RAN has been architected and designed, but for completeness, this topic is included. More likely than not, there is a high probability that the network operator has multiple bands of operation and is already deploying or considering deploying multiband antennas to efficiently use the tower space. The antennas are connected directly to the RUs, but in the case of a Brownfield operator who has legacy deployments, this antenna is shared by both the O-RAN Radio and legacy remote radio head (RRH), or in many cases, by multi-RAT radios, e.g. 4G and 5G.

As part of the antenna selection, one needs to consider the number of radio ports being connected and the mounting of the antenna and the radio on the tower. Compliance with a preagreed standard bracket design will greatly help in deployment. One of the most common deployment mistakes is cross cabling, especially in multiport antennas, i.e. connecting a cable to the wrong antenna or radio port, which is very prevalent when a lot of cables are involved. Methods such as coloring the cables, proper placement of the connectors, and using cluster cables can minimize the mistakes significantly. The jumper cable type and length have an impact on the link budget due to insertion losses. Low-loss jumper cables can be used, but they are usually very thick and become unwieldy for deployments with tight space restrictions and with multiple ports. Another method is to use a direct connection between the radio and antenna without any cables (the blind mate option). This deployment model minimizes the distance between the radio and the antennas to keep the insertion loss very low, helping with link budget and coverage. This, however, requires close collaboration between the radio and the antenna manufacturers. Operators also have to study the thermal characteristics of the radio and heat dissipation angles. There needs to be some space between the antenna and the radio for airflow, allowing the radio to dissipate the internal heat. A classic bookshelf deployment occupies the least space, but again it is important to consider thermal aspects and airflow to maintain good operating temperature. The radios tend to lower their transmit power by a few dBs at high temperatures to remain operational.

The link budget, structure type, local zoning regulations, and coverage target define the type of antenna used, but for some Brownfield operators, replacing the antenna may not be a viable option, and in such cases, the O-RAN components must fit within the existing framework. Obviously, the bigger the antenna, the more the gain and coverage area. However, other factors determine the limitations, such as wind loading or mounting on the side of the building that may restrict the antenna size. An operator always needs to have an array of antenna options including size, beamwidths, and gains, for the RF engineers to optimize the network.

It is very important to have clear antenna specifications, both electrical and mechanical. With cooperation from across the wireless industry, the Next Generation Mobile Network (NGMN) Alliance has developed a set of base station antenna standards to make antenna selection easier.

This is commonly known as BASTA standard (Standard Recommendation on Standards for Passive Base Station Antennas BASTA standard v12 Dec 23 2021), named after the working group that created the specifications.

Antennas also play an important role to enable coexistence between different operators who are co-located on the same cell site. 3GPP and the regulatory bodies may have specific protection and coexistence criteria defined for certain frequencies. Such consideration may include receiver protection in terms of Out of Band Emissions (OOBE), or blocking characteristics. That must be considered. For example, a high-power transmitter (TX) should not be located in the line-of-sight of a receiver (RX) that is receiving on the same band. This occurs mostly in time-division-duplex (TDD) scenarios where the sync is not maintained. Similarly, if a TX is adjacent to an RX with a minimal guard band in the frequency domain, this can result in receiver overload scenarios. External filters alone may not be enough to provide sufficient protection; other RF techniques including modification of the antenna location, adjusting beamwidths, horizontal azimuth change, and down tilts should be used to manage interference. For the most part, RF engineers focus on horizontal and vertical beamwidths, but for interference management purposes, one needs to focus on front-to-back ratio, sidelobe suppression, and interband isolation. Many times, the engineers take the most conservative approach and try to create a common solution, i.e. filtering for all possible scenarios, however remote they may be. This conservative approach increases the cost of the overall solution and the performance. Site-specific engineering should be used instead of a global one as the one-size-fits-all approach is not ideal.

10.2.1.2 O-RAN – Radio Unit

The O-RU, or simply RU, is at a basic level, a conduit between DU/CU and the end user device. An RU has two main components: the baseband board (BB) and the RF Front End (RFFE). The BB, the frequency-independent board, has the following key functions:

- evolved Common Public Radio Interface (eCPRI), Control User Synchronization and Management (CUSM) plane, Timing synchronization
- Low Physical Layer including Inverse Fast Fourier Transform (IFFT) and Fast Fourier Transform (FFT), cyclic prefix (CP), Physical Random Access Channel (PRACH) filtering, In-phase and Quadrature (IQ) handling
- Crest Factor Reduction (CFR)
- Digital Pre Distortion (DPD)

As with any new technology, and even for O-RAN, a typical implementation for the first two functions is with FPGAs. Now that the capability and interfaces have quite matured especially for Category A radios, the RU vendors are exploring to move toward ASICs that are faster and consume less power. The first two sets of functions are typically the IP of the chip provider who provides the software, drivers, and libraries to the vendors for implementation.

The BB controls the bandwidth, number of MIMO layers, duplexing, etc. A single BB can be connected to more than one RFFE. The eCPRI interface connects the Low PHY on the Radio with Hi-PHY of DU. Depending on the bandwidth, it can be a single 10G link or up to multiple 25G optical links. Another standard like Radio over Ethernet (RoE) is an alternative to eCPRI but has not seen wide-scale adoption by the industry. The eCPRI interface also controls the compression used for communication between the RU and DU – compression saves costs on SFPs. For multiradio deployments, one method is to have eCPRI cascading between the radios to save cost by running fewer optical cables, but the extra port can also be used for redundancy.

The RFFE contains frequency-specific subcomponents such as

- Filters
- Power Amplifiers

There can be many types of filters, but the most common ones are Cavity filters that provide the best rejection with the least insertion loss. Filter insertion loss is proportional to the bandwidth and the isolation required. At the same time, the size of the filter is inversely proportional to the frequency. Hence, the size of the filter for low band spectrum is larger than for the high band spectrum. The filter performance is not only dependent on regulatory requirements but is also dependent on the level of coordination and co-existence required. It is beneficial to have filtering built-in to the radio to avoid additional cabling, while being operationally efficient from a supply chain and build process standpoint. With in-built filters, the operators do not have to handle extra components for installation and inventory. However, adding internal filters means that all sites, whether filtering is needed or not, will have the same RU even if it is not needed. This can lead to additional unwanted costs.

The most common power amplifier type used in radios is a Doherty Amplifier. The PAs usually run with an efficiency factor of 30–40%. Higher wattage PAs require larger heat sinks. The newer-generation PAs are GaN (Amplifiers)-based and more efficient. It is important to appropriately dimension the radio capability and not always go for the highest power ones, as for the most part, the system is uplink limited. What is seen time and again is that operators define the Radio with max power for flexibility but in most cases, end up operating at lower power to manage interference as the inter-site distances shrink.

Most radios utilize passive convection cooling, but active cooling with fans is also an option where size is a restriction. In the past, the reliability of a fan-based solution for heat dissipation was a concern. Specifically, there is a risk for fans to go bad, and these were not field replaceable. Hence, the entire unit had to be replaced. The higher return rate resulting in a higher cost to both manufacturer and operator was a big negative for active cooling. However, the newer-generation components are now exhibiting higher reliability. Operators should always consider the current state of the technology for this decision. Another such technology is liquid cooling. Liquid cooling technology is used within the data centers to carry heat generated by the servers and networking gear and provide good thermal dissipation. Similar technology is also employed by some radio solutions in the market achieving excellent results. Also to clarify, radio thermal dissipation design is agnostic of O-RAN or traditional deployment, but with O-RAN, operators do get a wider choice of vendors and radio solutions. The operators have options to explore new technologies and make selections instead of being handed down a solution. It is no longer a closed black box.

On the amplifier itself, usually one specifies the average output power. However, due to the nature of orthogonal frequency division multiplexing (OFDM), the PA has to manage a high peak-to-average power ratio (PAPR). The peaks can be as high as 10–13 dB. For example, for a PA with 40 W capability, the PA needs to manage peaks of 400–800 W, otherwise, the CFR will clip the incoming signal resulting in error vector magnitude (EVM) degradation. The system should be carefully designed to handle PAPR both at the input of CFR and DPD to have the least EVM and lower block error rate (BLER) in the modulation constellation.

On the receiver side, the most important characteristic is receiver sensitivity. Receiver sensitivity is defined by the target Uplink Data Rate which is a function of target cell edge design and can change widely. Noise Figure (NF) is one of the key RU variables that impact receiver performance. A general rule of thumb is to have 2.5 dB NF for FR1 FDD radios and 3 dB for FR1 TDD radios at nominal operating temperatures, with a margin of 0.5 dB for extreme temperatures. The choice of internal filters, and other components such as circulators or switches can impact the NF too. Another practical characteristic for the Receiver is the Adjacent Channel Selectivity (ACS), i.e. how the receiver can decode the wanted signal in presence of high interference in the adjacent channel. This scenario is very prevalent in cellular systems and is identified as a near-far problem. Minimal ACS performance is described in 3GPP (3GPP TS 38.104 2022); this is exceeded by almost everyone by several dBs. Operators should study this in detail and specify it as part of technical requirements and use it for vendor solution selection.

10.2.1.3 O-RAN-Distributed Unit (O-DU)

O-DU, or DU as commonly known, is the entity that handles the high split of the Physical Layer (Hi-Phy), Media Access Control (MAC), and Radio Link Control (RLC). MAC and RLC functionality are well defined in detail in 3GPP (3GPP TS 38.300 2022) and remain the same for both the traditional and O-RAN-based solutions.

In the case of 7-2x split, the interoperability testing (IOT) profile is negotiated between the DU and RU vendors, considering aspects such as compression, M-Plane, number of cells, etc. Currently, most of the DU functionality is implemented on x86 processors and uses either an FPGA or an eASIC-based accelerator to handle compute-intensive calculations including Forward Error Correction (FEC). There are other options too such as GPU, in-line accelerator cards that are coming into the market as alternatives but are not mature yet at the time of writing this chapter.

Multiple RUs are connected to a single DU. However, a single RU cannot connect to more than one DU unless implemented in RAN sharing mode. The specifications for such implementation are still in works within the O-RAN standards at the time of writing this chapter.

10.2.1.4 O-RAN-Centralized Unit (O-CU)

O-CU, also referred to as CU, handles the radio resource management (RRM) and PDCP layer of the RAN. Similar to the DU, 3GPP has defined CU functionality in detail and it remains the same between a traditional RAN and an O-RAN.

Unlike DU, CU is further split (3GPP TR 38.806 2017) between CU-CP (control plane) and CU-UP (user plane). A single DU is mapped to a single CU-CP but can be mapped to more than one CU-UP. Accordingly, CU-CP can manage multiple CU-UPs but a single CU-UP is anchored to a single CU-CP only. This architectural approach makes complete sense as it allows for independent scaling and optimal resource utilization since the user plane is more compute-intensive than the control plane, consuming significantly more resources.

CU-CP is limited by the number of cells it supports and handles all call processing between users and the Access and Mobility Management Function (AMF)/Mobility Management Entity (MME) in the core network. CU-UP is defined by its packet processing capabilities (packets per second – PPS), which is also a function of the underlying compute. CU-UP is connected to the UPF or Serving/Packet Data Network (PDN) Gateway (S/P-GW)-U. As mentioned earlier, each CU-CP can have multiple CU-UPs connected to it, in the case of slicing one may even assign each user bearer or PDN to separate CU-UP that may or may not be collocated with each other.

10.2.1.5 RAN Intelligent Controller (RIC)

RIC is perhaps the most powerful and impactful function of the O-RAN architecture. Broadly, it consists of two parts – the near-real-time RIC (near-RT RIC) and non-real-time RIC (non-RT RIC). Though architecture is nearly complete within the O-RAN standards (O-RAN.WG1.O-RAN-Architecture-Description-v06.00 2022) but much work remains on the use cases and supported messages. Both RIC types require a platform to run their applications (RICAPP) known as xApps for near-RT-RIC and rApps for non-RT-RIC. Many of the commonly envisioned rApps are akin to self-optimizing networks (SON) functionality. xApps operate at a decision time scale of 10–100 ms, while rApps operate at >1 s time scale.

RIC deployment is optional and does not qualify a network as O-RAN compliant or not. Of the interfaces defined for RIC, E2 (O-RAN.WG3.E2SM-RC-v01.03 2022) is critical. There is still work required for intervendor interoperability at the time of writing this chapter. Another aspect that is important for RIC is the simplicity for application providers to create apps on the platforms using SDKs and APIs. Ideally an operator would want 100s of applications to choose the right ones. Other requirements regarding Apps life cycle management (LCM), high availability of the platform, interapp communications and data management for Apps are still being worked out by O-RAN standards (O-RAN.WG3.RICARCH-v02.01 2022) and the industry in general.

10.2.2 Traditional vs. O-RAN Deployment

There are fundamental differences between the deployment models of conventional RAN, vRAN, and disaggregated O-RAN. In the conventional mobile network, all RAN components are predominantly deployed at the cell site, that is, antenna, RU, baseband unit (DU + CU) are installed at the cell site. In case of vRAN, the DU and CU are virtualized but do not necessarily have open interfaces between them. In contrast, the disaggregated virtualized RAN has a decoupled baseband and radios usually based on option 7-2x (intra PHY split) and option 2 (DU-CU split), the CU is further split into CU-CP (control plane) and CU-UP (user plane). Traditional RAN and vRAN can only be deployed on-prem while O-RAN can be deployed in a hybrid cloud. Figure 10.2 illustrates the difference between the three deployments.

The RAN software is implemented on either custom ASICs or General Purpose Processors (GPPs) to pool and scale the computational resources. Other deployment options, such as integrated RU and DU with disaggregated CU and integrated gNB are also possible and can be considered as O-RAN compliant as long as the interfaces remain compliant irrespective of the placement of the function itself.

The Antenna and RU are deployed at a cell site. The DU may also be deployed at the cell site or deployed in a data center-type setting with pooling since a single DU can manage more than one site as illustrated in Figure 10.3. Similarly, CUs (CU-CP and CU-UP) can be deployed in the data center providing aggregation gains. From a hierarchy perspective, CUs are deployed at a higher level, the CU-CP or CU-UP may be deployed in the same or even different data centers leveraging Control and User Plane Separation (CUPS) architecture. The disaggregated architecture provides options to operators to mix and match, i.e. DU from Vendor A, CU-CP from Vendor B, and CU-UP

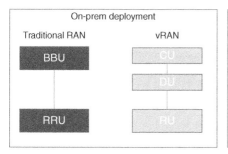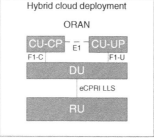

Figure 10.2 Traditional vs. vRAN vs. ORAN.

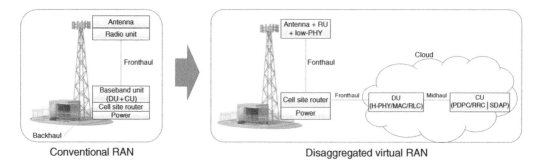

Figure 10.3 Disaggregated vRAN vs. conventional RAN.

from Vendor C. Of course, this increases integration complexity and operational burden of managing multiple vendors, but nevertheless that is a flexibility offered by an open architecture.

10.2.3 Typical O-RAN Deployment

Usually, each cell site consists of three sectors, and each sector has an antenna connected to one or more RUs, as illustrated in Figure 10.4. This antenna and RU combination is dependent on the frequency of operation, bandwidth, and connectors. Nowadays, there is a wide availability of multiband antennas allowing operators to connect multiple radios to a single antenna. Further, Radio designs have evolved to support multiple bands within the same enclosure.

In the network design, it is important to pay attention to fronthaul data rate and to have the right optical ports on both ends. The fronthaul data rate is a function of the carrier bandwidths, number of carriers, MIMO layers, and compression. Note: A "no compression" option leads to a higher fronthaul data rate but could result in compute savings.

There is also consideration of the type of optics used, i.e. there is a wide range of options available like short range, long range, colored, etc. The choice is dependent on the fronthaul distance and the type of optical gear.

The F1 connection between DU and CU is also referred to as the midhaul. The midhaul needs to meet certain latency depending on the use cases offered and the bandwidth which is directly proportional to the RF bandwidths, modulation, and the MIMO layer. However, in the midhaul case, unlike the fronthaul, one may get the benefit of aggregation. As all cells in the site do not peak at the same time, statistical benefit allows for provisioning the midhaul link at a lower rate. The required data rate of each connection is dependent on the bandwidth, the number of cells, and the number of carriers.

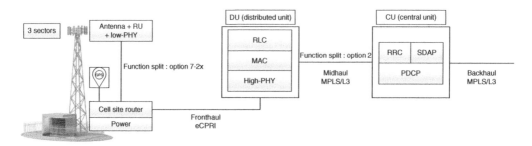

Figure 10.4 Typical O-RAN macro deployment.

The network equipment processing delay and the compute processing delay should be considered in defining latency, this is more important for a Cloud RAN (C-RAN)-type deployment. Typically, the max fronthaul delay is referenced to be about 160 μs (O-RAN-WG4.MP.0-v09.00 2022), but this can be extended to 200+ μs for certain use cases by adjusting the buffers and by reducing the processing time on either end. This has a direct impact on the radius between cell site and the DU and therefore more sites can be aggregated at a single location.

The transport protocol between DU and RU is L2 eCPRI but between CU and DU, it can be L3 MPLS (Multiprotocol Label Switching) or others.

10.2.4 Spectrum and Regulatory

There are no special restrictions on what frequencies can be used for O-RAN deployments, i.e. regulatory requirements are the same between traditional and O-RAN. Both FR1 and FR2 can be used for O-RAN with no restrictions and similarly the profile is defined for both FDD and TDD. Certain narrowband technologies like NB-IoT and eMTC are being defined in O-RAN specifications as extensions. Similarly, there are no regulatory requirements for O-RAN. O-RAN-based networks shall also support all regulatory requirements for Voice, CALEA, Legal Intercept (LI) that are supported by traditional networks.

10.3 Network Planning and Design

Standard RAN features and functionalities are governed by 3GPP releases, and Open RAN does not change that. As part of a release planning exercise, both vendor and operator need to align and select which features will be delivered, tested, and deployed in the network. There is also a huge dependency on the user equipment (UE) ecosystem that drives the priority of those features. IOT is performed between device chipsets and RAN vendors to validate 3GPP compliance. From a UE perspective, a UE does not know or even needs to know whether it is working on an O-RAN or traditional RAN.

Certain features such as Coordinated Multipoint, traffic management, RIC, etc. have an impact on the network design such as delay, data flow, capacity dimensioning, and transport (*O-RAN. WG3.E2SM-RC-v01.03*). The architecture has to be flexible enough to accommodate all scenarios. Although O-RAN provides tremendous flexibility to deploy the network, there are few basic tasks and prep work that are similar for any radio deployment that needs to be planned ahead. Some of them are covered here:

- Identify the target Area of Interest (AOI) that needs to be covered by the network.
- Pin down the target services that need to be offered which drives the link budget and spectrum usage. Link budget is used to create a detailed view of system capability considering the coverage objectives that include area coverage probability and cell edge reliability of the design, in building coverage, body losses, etc.
- Know the available sites within the AOIs. For a greenfield deployment, the RF teams have to run several iterations to derive an optimal combination of sites with available heights for antenna placement. For Brownfield, it may either be replacement or augmentation, but nevertheless, there needs to be a baseline of sites for the RF team to start with.
- Identify any spectrum coordination that is required, either co-existence or other regulatory requirements that are applicable to the AOIs.

- RF design using the link budget and tuned propagation model is an iterative process with specific models applied to a particular morphology and topology. The complexity increases with a multi-band deployment. The Optimization Engineers have to approach with a layer cake coverage and perform proper mobility management for seamless user operation while maintaining maximum spectral efficiency per band.
- Equally important is the availability of the right transport at the site. Plan ahead your transport providers and the target IP address schemes.
- Determine the available data centers, rack space, and port availability for interconnect. Always plan for additional racks within the data centers for future expansion and prefer to have racks adjacent to each other to avoid running long cables across the data center.
- Finally, create a detailed BOM and calculate the power consumption of the equipment at the site and data centers, clearly identify the required cabling and cabinet sizing.
- Apply for permits and conduct civil works as needed.

Proper planning is more critical for an O-RAN-based network compared to a traditional network due to the disaggregated architecture. Networking and IP are very critical and need to be done right. More details are provided as follows.

10.3.1 Cell Site Design

A cell site usually consists of two parts, elements that are mounted on the top (i.e. radios and the antennas) and the elements that are placed at the foot of the structure. Usually, there is a temperature-controlled room or cabinet that houses the electronics required to manage, operate, and connect that particular cell site.

Some common equipment includes a Cell Site Router (CSR), DU compute, Batteries, Rectifiers, and Out-of-Band Management console. The batteries are used for power backup during an outage. These batteries especially consume a lot of real estate and are very expensive, and need to be carefully planned on how and when they will be used. DU compute configuration is covered in more detail later in this chapter but it should be noted that with the virtualized approach, the compute servers have to be specified such that it has the right depth and can operate in the operating temperature of the cabinet or the housing. A regular data center server cannot operate at a cell site.

The Cell Site Router (CSR) should be eCPRI-capable, GPS-enabled, and be timing-aware acting as a boundary clock and have a sufficient number of SFP ports to support the connectivity to not only the radios but also to the DU compute(s) and other components in the rack or the cabinet. With virtualized DU, separate ports may be required for operation, administration, and maintenance (OAM), precision time protocol (PTP), Kubernetes (K8S) management, BIOS management, etc., in addition to the fronthaul and midhaul connectivity. It is prudent to have spare ports available in the CSR in case of a port failure, as it is much cheaper to use another port on the same device instead of replacing the whole unit. The backplane of the CSR should be sufficient to handle all the bidirectional traffic, especially the fronthaul traffic which is large. Since the CSR is likely responsible for providing timing and sync, most likely it will be a lower-layer-split (LLS)-C3 configuration that is described later in this chapter. If the deployment is at smaller sites with limited radios, one may not need a CSR and connect Radios directly to the DU in an LLS-C2 configuration, but this not only requires the compute to have a sufficient number of ports but also have a routing application that can handle layer 3 and layer 2.

Other components such as batteries, rectifiers, management consoles, etc., are the same for both O-RAN and traditional cell sites. It is recommended to have spare cables in the conduit from the base of the tower to the radio both for power and fiber as this will not only help during the expansion but also acts as a failover.

10.3.2   Network Function Placement

Almost all RUs are RRHs, i.e. they are deployed on the top of the tower or the building and very close to the antenna. Radios are no longer deployed on the ground with a Tower Top Amplifier (TTLNA).

The compute hosting the DU application can either be placed at the cell site or in a different location but within the fronthaul latency budget. The choice of the DU placement is a function of transport availability, anticipated pooling gain, number of cells, and other operational factors. Irrespective of whether the DU is at the site or in a remote data center, it can still be implemented with the same O-RAN split architecture.

The CU placement has more flexibility due to the higher midhaul latency budget. The distance between the CU and DU can be several milliseconds but it is preferable to keep it to low single-digit ms latency to have low RTT delays that may be needed for certain applications. The CU can be placed either in Public Cloud or On-Premises or both. This can be further separated by having CU-CP in the Public Cloud and CU-UP deployed on-prem.

10.3.3   Dimensioning

Unlike traditional black box RAN solutions that are over-dimensioned and one-size-fits-all units, with O-RAN, operators have an option to right size the solution and control the cost. But, this requires proper planning and understanding of the solution. We can start with CU and DU application dimensioning, which are a function of spectrum bandwidth, throughput, number of users, number of cells, etc., which is a good start for network dimensioning but that alone does not provide the full picture. One also has to think from a higher level such as market level or cluster level, etc. With the centralized and virtualized nature of the development, we have to consider both horizontal and vertical scaling of the application along with statistical gain achieved by multiplexing at different levels.

More importantly, with 5G, the operator has to consider the impact on application performance since the dominant application may no longer be a voice. With the growth of over-the-top (OTT) applications for communications, the ability to move bits from inside to outside the network with minimal latency is key. The network deployment is also driven largely by use cases. 5G with a 1 ms latency requirement is applicable to private networks with edge compute combined with a CU and DU. This is not the same as a macro cellular-type deployment.

Another consideration that operators typically use for dimensioning is the projected future growth, as the user base grows, and as the usage per user increases year over year, additional spectrum introduction, small cells, etc., all these impact the network dimensioning. But a multi-year projection is rarely right. Open-RAN provides the ability to scale up in modular step size instead of buying infra for the projected end state. The benefit of an open virtualized network is that it can scale based on demand, e.g. operators can have multiple CU-UP NFs spun up as demand increases during, e.g. a gaming event, and scale down when the event is over. This level of flexibility when combined with CD pipelines and network observability are pretty important tools for an operator to dynamically adjust the network.

This section provides details on different considerations for such dimensioning.

The dimensioning can be broadly divided into two subparts:

1) Application dimensioning
2) Platform dimensioning

10.3.3.1   Application Dimensioning

Starting with the DU, the first and foremost thing to consider for dimensioning is the DL and UL bandwidth (MHz) and the number of sector carriers (Cells/DU). Since most of the currently known O-RAN DUs and CUs are based on Intel x86 processors, this will be used as an example but the logic can be extended to other compute types too. Note, uplink consumes significantly (~50%) more compute resources compared to downlink as additional decoding processing is required. This multiplied by the number of MIMO layers on both uplink and downlink will directly impact the compute and memory requirements. Other aspects such as the type of Fronthaul Compression used impacts the compute, more compute is required for higher compression levels. Along similar lines, usage of Jumbo Frames can impact compute too since additional time would be needed to process the entire packet and buffering aspects would need to be considered.

For CU, separate dimensioning for CU-CP and CU-UP should be performed. CU-CP dimensioning is based on the operator call model which directly defines the signaling load, user capacity and the number of cells. Typically, the CU-CP can support 256–2048 cells per instance; this provides a lot of multiplexing gains. There is a strict mapping between Cells and CU-CP, and it is recommended to map all cells in close proximity to each other to the same CU-CP to avoid inter-gNB/eNB handovers. For this reason, always plan and leave some capacity for future growth i.e. addition of more sites or cells on the existing ones. If the CU-CP is implemented using cloud-native principles, it can scale horizontally as the operator builds more cells. For CU-UP, the dimensioning is more or less dependent on the number of packets it has to process. Typically, each CU-UP has a certain PPS limit and if the CU-UP is cloud-native, it would scale horizontally adding more pods with a capacity increase. But there is a limit to horizontal scaling, after which the operator has to plan for newer instances of the NF. The max offered speeds are also dependent on the underlying network interface card (NIC) on the compute node. An application might be capable of processing 100G traffic but if it uses only a 25G NIC card, that can limit the maximum offered capacity.

10.3.3.2 Platform Dimensioning

Some platform components consume compute resources that need to be accounted for. If a hypervisor is used, overhead for the hypervisor and similarly the compute requirements of the operating system (OS) needs to be added. Other lesser known but equally important overhead is compute required for OAM to collect performance and fault metrics of not only the applications but also the platform and the infrastructure itself. Lastly, if the applications are containerized (e.g. Kubernetes), they need to have compute for running the Kubernetes worker node and, in the case of discrete clusters, have to include the overhead of the Master Node too.

Typically, Hyper-Threading (HT) is enabled on the x86 processors. So for every physical core (pcore), there are two virtual cores (vcores). HT does not double the processing capacity but does provide a significant performance boost. None of these platform overheads are heavy and consume only one or two vCores each but they do add up and more importantly, these compute cores are not available for the applications to use. In addition, PTP support may need to be accounted for in the DU as overhead.

The final thought on the dimensioning – the above compute allocation is for basic operations, but to make a system resilient, to minimize downtime, to manage upgrades and updates, and to ensure high availability, additional overheads like multiple pods and redundant microservices should be added. This will require additional compute and storage capacity. The operator should spend time with the O-RAN vendor to understand their architecture and design, before making the infrastructure selection.

10.3.4 Virtualization Impact

O-RAN deployments can take advantage of decoupling of software from hardware that started with virtualization, which was utilized to a limited extent in 4G/LTE. NFs implemented in containerized packages allow the deployment of 5G NFs on commodity hardware, reducing the cost of network deployment for operators significantly. Additionally, containerization technologies allow multiple NFs to be deployed simultaneously on shared compute, network, and storage resources. Containerization also provides inherent resilience due to the ability to re-instantiate malfunctioning pods very quickly and scale NFs to meet increasing demand. Virtualized NFs also enable slicing by allowing multiple instances of NFs to be running simultaneously to meet specific customer needs.

Here is a short description of some of the resources and configurations used by the virtualization platform to enable Containerized NFs on standard x86-based COTS hardware. Any combination of the resources and configurations may be used on the underlying hardware and the virtualization stack to meet the specific requirements of the containerized NFs (CNFs) that are to be instantiated on the platform. Some of these requirements are independent of whether the underlying infrastructure is on-prem (on-premise) or a public cloud provider (PCP)'s infrastructure. The operator has to verify support for these requirements before deployment. Many of these requirements are described by using Intel x86 architecture as an example and are critical for RAN operations.

10.3.4.1 Non-Uniform Memory Access

In a multi-core server, such as a modern Intel Architecture (IA) server, some areas of memory and input/output (I/O) devices are closer and faster to access from some CPU cores than others. By grouping similar cores together as a non-uniform memory access (NUMA) node, and allowing them to use only the memory that is "local" to them, memory access times can be reduced. Such optimization is beneficial for applications such as Network Functions Virtualization (NFV). Proper NUMA settings ensure that code is adjacent to the CPU core that executes it, resulting in significant performance improvement. Typically, in a two-socket system, each socket would correspond to a NUMA node. For high-performance applications like DU, NUMA boundaries should not be crossed.

10.3.4.2 Hyper-Threading

HT is Intel's proprietary Simultaneous Multithreading (SMT) implementation used to improve parallelization on their CPUs. HT may be enabled to improve the performance of multithreaded applications and disabled in intense computing (single-threaded) environments.

The current version of Intel's HT technology allows a maximum of two threads to be dispatched to an HT-enabled CPU. To the OS, an HT-enabled CPU simply looks like two processors and thus the OS sends two threads to the CPU for execution.

10.3.4.3 CPU Pinning

CPU-pinning is a means of ensuring that a virtual CPU (vCPU) in a particular VM/pods always runs on a dedicated physical CPU (pCPU). In combination with CPU thread isolation, this will prevent the CPU core from being shared with any other VM/pod, and is important for applications with high performance and low latency requirements.

10.3.4.4 Huge Page

The addresses of memory locations available to a process in Linux need to be mapped to the actual physical memory locations. Rather than map individual bytes of memory, contiguous regions of memory called pages are used. The default size of pages is 4 KB, and in modern systems that have several hundred GB of RAM, the total number of pages becomes very large. This causes the address translation to become slower, as it becomes more likely there will be a miss on the cache used in this process, called a Translation Lookaside Buffer (TLB). Linux offers larger page sizes of 2 MB or

1 GB to reduce the probability of TLB misses. Such pages (minimum unit of memory allocation) are called Huge Pages and will improve the performance of NFs.

10.3.4.5 Single Root Input/Output Virtualization

The Single Root I/O Virtualization (SR-IOV) interface is an extension to the PCI Express (PCIe) specification and is required to achieve performance that is nearly the same as in non-virtualized environments. The SR-IOV port from each NIC can be used for SR-IOV traffic. Each SR-IOV port can have a maximum of 32 virtual functions (VFs) (depending on the NIC type) and the number of VFs to be created is configurable. SR-IOV enables network traffic to bypass the software switch layer of the virtualization stack, the network traffic flows directly to the VF, and child partition.

10.3.4.6 PCI Passthrough

Peripheral Component Interconnect (PCI) passthrough allows a VF to have full access and direct control of physical PCI devices on the host for a range of tasks. PCI passthrough allows PCI devices to appear and behave as if they were physically attached to the guest OS. Combined with SR-IOV, a physical device is virtualized and shows as multiple PCI devices. Virtual PCI devices can be assigned to the same or different VFs. In the case of PCI passthrough, the physical device cannot be shared and is assigned to only one VF. This is needed to have prioritization and separation of traffic.

10.3.4.7 Data Plane Development Kit

Data Plane Development Kit (DPDK) is a framework composed of libraries and drivers that improves packet processing performance and throughput. DPDK makes use of different techniques to accelerate packet processing such as Kernel bypass creating a path from the NIC to the application; Poll Mode Driver (PMD) where the CPU polls the NIC for new packets instead of the NIC interrupting the CPU when a frame is received. DPDK also includes the wireless baseband library (BBDEV), which provides a common programming framework that abstracts HW accelerators based on FPGA and/or Fixed Function Accelerators that assist with 3GPP Physical Layer processing. Furthermore, it decouples the application from the compute-intensive wireless functions by abstracting their optimized libraries to appear as virtual BBDEV devices. BBDEV is part of the O-RAN architecture.

10.3.4.8 Resource Director Technology

Resource Director Technology (RDT) provides a framework with several component features for cache and memory monitoring and allocation capabilities. These technologies enable tracking and control of shared resources, such as the Last Level Cache (LLC) and main memory (DRAM) bandwidth, in use by many applications, containers or VMs running on the platform concurrently. RDT also aids in "noisy neighbor" detection and helps reduce performance interference, ensuring the performance of key workloads in complex environments.

10.3.4.9 Cache Allocation Technology

Cache Allocation Technology (CAT) helps in quieting the noisy neighbor and allocator proper resources to apps or VMs that have higher priority. CAT provides software-programmable control over the amount of cache space that can be consumed by a given thread, app, VM, or container. It is used in quieting the noisy neighbor and allocator proper resources to apps or VMs that have higher priority. This allows, for example, OSs to protect important processes, or hypervisors to prioritize important VMs even in a noisy data center environment.

10.3.4.10   Resource Overcommitment

Hypervisors allow CPU and RAM to be overcommitted for instances, which means that more resources may be allocated for instances than are actually present physically. The hypervisor takes care of scheduling the instances in a way that it appears that they have dedicated resources. This will cause some performance degradation, which may or may not be acceptable based on a particular VNF. The factor by which resources are overcommitted is called the overcommit ratio.

For example, in a two-socket server that has 20 pcores per socket, there are a total of 40 cores. With HT enabled, there will be 80 vcores. With a CPU overcommit ratio of 10, there are 800 virtual cores that may be assigned to instances. (The actual number available will be less, because some cores will be used for other purposes, such as running the host OS, and the virtual switch.)

10.3.4.11   Operating System

An optimized real-time distribution of the Linux operating system (OS) with a light footprint is needed for running delay-sensitive applications, especially the DU. This OS contains container runtime daemons and works with container orchestration frameworks such as Kubernetes. Real-time optimizations are needed to meet the latency and performance requirements necessary for the RAN workloads.

10.3.4.12   K8S Impact

Operators looking for maximum flexibility may adopt multivendor, multicloud approach and the use of containers to virtualize the various NFs that comprise the Core, the RAN, and the management layer of the wireless network. The use of Kubernetes to deploy, run, and manage containerized applications in distributed, multicloud environments has to be carefully planned.

Kubernetes construct has two main parts, K8S master and K8S Worker. The K8S Master is designed to manage multiple workers which are typically a compute node each. Usually, the Master and Worker nodes are in separate nodes, e.g. in public clouds, but there is also an option for both Master and Worker to share the same node, in such a case, the Master is only managing that particular Worker only. There are advantages to both options and the operator may use both deployment modes application by application. From a dimensioning perspective, one has to account for the overhead for Worker and Master. For a remote Worker deployment, there is a strict round-trip latency from the Master that has to be accounted for in the network design. Although the Worker node can continue its operation in case of Master failure, LCM breaks down until the Master is back in operation, so Masters are deployed with high availability design. The mapping between Worker and Master nodes has to be carefully designed for O-RAN, especially for deployments where DU is at the site, there is a tradeoff between efficiency and failure domain.

10.3.5   Networking Hardware

It is preferred for the networking equipment to have an integrated GNSS receiver and to offer precise frequency and phase/time synchronization using the latest industry standards, i.e. equipment should be providing PTP Synchronous Ethernet (SyncE)/1588 Class B timing. If features like MACsec are enabled, we need to ensure that the line rate can handle the additional overhead. One should look at the power consumption per bit as it quickly adds to the OPEX and also capabilities like deep packet buffering, and optimized forwarding architecture.

Operators can also get white-labeled hardware and add a third-party virtualized routing and forwarding software. Usually, the hardware in such cases is based on merchant silicon such as Broadcom ASIC. In such cases, the operator is responsible for integration, feature management, and richness of the telemetry information. It is recommended that software support standard network configuration protocols such as Netconf and Yang (YANG).

For O-RAN, networking equipment should support at least Layer 2, Layer 3, and MPLS protocol. The operator can also use Segment Routing (SR) with MPLS (SR-MPLS) or with IPv6 (SRv6). To maintain end-to-end QoS, the networking gear should be QoS-aware and able to apply the prioritization rules with the right security construct. Always ensure that there are sufficient and extra ports on the hardware, supporting QSFP and QSFP+, depending on the deployment scenarios you may need multiple 10G, 25G, and 100G ports. A hot-swappable power supply with $1+1$ redundancy is recommended. As one moves up in the architecture, the networking gear should also be more powerful and handle multiple 100GE, 200GE interfaces, and even 400G/1000G interfaces too.

All routing or switching elements that are in the O-RAN fronthaul path between RU and the DU need to adhere to a stringent set of requirements in the front-haul data path between the RU and the DU. A high-performing router platform must be selected that combines high throughput and low latency – two key requirements of any front-haul data plane element. Ideally, switching latency should be less than 2.5 µs with a switching jitter of less than 5 ns.

Other factors impacting the total latency of all routers and switching platforms include:

- **Serialization delay:** This is a function of packet size and interface speed. The larger the packet size, the longer it takes to transmit that packet as a linear series of bits out of a given interface. The higher the interface speed, the lower the amount of time will be required to transmit that packet as a linear series of bits.
- **Queueing delays:** If an interface becomes congested, some amount of queueing will occur, even if some packets are intentionally dropped to ease the congestion. Oversubscription of network resources should be avoided if the network is intended to support latency-critical applications.

10.3.6 Hardware Type

It is very important to select the right hardware type for telco deployment, standard COTS server cannot be installed at the cell sites and telco data centers, e.g. Telco Data Centers (DCs) require Network Equipment Building Systems (NEBS Requirements) compliance that standards COTS servers do not qualify for. There are different levels of NEBS compliance depending on the location of DC, this adds to the cost of the server too. Secondly, the cell site cabinets usually aren't the same depth as telco racks, depending on the cabinet design one may require more access to the front of the cabinet. Some of the factors for any telco gear are that they be temperature-hardened, small form factor, and have low power consumption. Another aspect that is often overlooked is security, you need not only hardware security but also software protection.

Typical networking products are 1RU high with a depth smaller than 300 mm with either side-to-side or front-to-back airflow. The operator should plan for typical, minimum, and maximum power consumption of all equipment and plan the power source accordingly, and also leave headroom for future expansions.

10.3.7 Reliability and Availability

Telco networks have always been designed with a target of 99.999% network availability, this resulted in every network component being extremely reliable with almost zero failure and this was made possible by adding redundant components resulting in a very high cost for the system.

However, in the virtualized space, the applications and the underlying general-purpose infrastructure cannot offer the same reliability as the purpose-built telco-grade systems. Hence, the general-purpose compute cannot meet the reliability metrics of a telco SoC. But at the same, it has also been very clear that many services can be offered by not adhering to strict telco availability metrics.

In its simplest term, an Operator should design a network based on the services and the service-level agreements (SLAs) being offered. An SLA is an agreement based on measurable metrics like uptime, responsiveness, and responsibilities. They are very hard to measure and also have a monetary impact if not met. Engineers end up spending a lot of time developing KPIs for such SLAs which end up being not used as much. Instead, these SLAs should be converted to SLOs, Service Level Objectives, for the design team to meet, e.g. the metric could be something simple like 99.999% of the packets will be transferred between Point A to Point B within 5 ms with 0.0001% packet drops.

Similarly, availability for virtualized RAN is a function of the availability of the application, platform, compute infrastructure, and connectivity. Due to the disaggregated nature of the deployment, we need to consider the availability of each component and the corresponding impact. Starting from the top, there is no redundancy for the antennas, which are made of passive components. Antennas rarely fail unless there is environmental damage due to hail or water leakage, in such scenarios the only recourse is to climb the tower and replace the damaged antenna. The RUs though have active components that are built for very large mean time between failures (MTBFs), i.e. they have very high availability. Similarly, other passive components like cables, optics, and connectors, all have long life unless improperly installed. All these can be categorized into a category of highly available but nonredundant components.

The second set of components belongs to the category of good availability but can be redundant. The examples of such components are cell site router, compute hardware (servers), Top of Rack (ToR) switches, etc. These components can be deployed in active: active or active: standby mode to improve system reliability but at the expense of additional Capex.

The third set is the collection of functions running as virtualized functions, such as DU, CU-CP, CU-UP, etc. These virtualized applications are a composition of many smaller modules of software, each module with a specific purpose such as data handling, error management, OAM, signaling, etc. Loss of certain sub function may not be service-impacting but it is operational impacting. The subfunctions that are service-impacting can be deployed with backup replicas or active: standby mode, resulting in incremental compute resources, while others can be deployed as singular units. These functions especially if run as microservices within a containerized solution are ephemeral. They are constantly orchestrated to scale and reproduce as and when needed. Such functions can also be deployed completely in Active: Standby or Active: Active mode.

10.3.8 Impact of Network Slicing

5G enables operators to offer network slices by allowing multiple logical networks to simultaneously run on top of a shared physical network infrastructure. This becomes integral to 5G architecture by creating end-to-end virtual networks that include both networking and storage functions. Operators can effectively manage diverse 5G use cases with differing throughput, latency, and availability demands by partitioning network resources to multiple users or tenants. Network slicing becomes extremely useful for applications like the Internet of Things (IoT) where the number of users may be extremely high, but the overall bandwidth demand is low. Each 5G vertical can have its own requirements, driving design considerations for 5G network architecture. Costs, resource management, and flexibility of network configurations can all be optimized with this level of customization. In addition, network slicing enables expedited trials for potential new 5G services and quicker time-to-market.

With Open RAN enabling disaggregation, slicing can be achieved at the NF level too, i.e. either operator can create slices within existing NFs or have functions dedicated to each slice and place them accordingly. For example, for a low latency slice, the CU and DU are located at the edge but for an eMBB slice, the CU can be in a remote data center while DU is at the edge. And with virtualization, the operator can instantiate and have the flexibility to move NFs on demand, i.e. the operator can enable a low-latency slice by instantiating a new CU at the edge and removing it when it's no longer needed.

10.4 Network Deployment

With Open RAN, there is more flexibility for the operators to deploy the network as per their needs. Here are some considerations for the deployment of each of the major NFs.

10.4.1 DU Deployment

10.4.1.1 DU Deployed at a Centralized Data Center

In this scenario, the DU is deployed at a centralized location such as a data center as illustrated in Figure 10.5. CU can also be collocated in the same data center but since the CU sits higher in the deployment hierarchy, it could be deployed in the same data center or a different one, as shown in Figure 10.5.

In this deployment, based on the O-RAN 7.2x split, the vDU is in a data center where multiple sites connect to it via dark fiber with low latency and sufficient bandwidth. There is a Top of Rack (ToR) switch at the data center which can take timing via PTP GM (or has in-built GM capability) and distributes timing to the radios and the vDU (LLS C3 configuration in O-RAN). To maintain

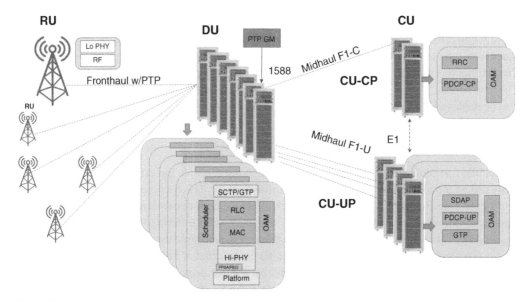

Figure 10.5 Centralized DU deployment.

timing and synchronization, the compute servers that hold vDU applications should have NIC cards that can not only support timing and synchronization with high accuracy that is needed but also have SyncE support for optimal operation.

The data center-type deployment allows for $N+K$ redundant compute architecture for high availability. Similarly, the ToR may also be deployed in $N+1$ fashion. It is preferred to have GPS on-site, but if not available, the GPS on the cell site can be used and the timing will have to be routed via PTP to the ToR switch. The latter approach requires GPS in more than one cell site to improve accuracy. It is recommended to have at least three GPS references to get the best timing using the Best Master Clock Algorithms (BMCA).

For the Midhaul, the DU Data Center Provider Edge (PE) router interface will terminate to a Network Interface Device (NID), which, in turn, connects to the Service Provider network to provide Ethernet Virtual Circuits (EVCs). Each EVC usually is switched in the Service Provider leveraging diverse paths avoiding any shared risk links or devices for resiliency. Each EVC, in turn, will terminate at the Data Center PE Router hosting the CU.

In this two-level data center hierarchy, the distance between the RU and the Data Center hosting DU can be up to 10–20 km depending on the fiber turns. It is really defined by the latency budget and the buffers within RU and DU. The connectivity between sites and the DU is using dark fiber to meet stringent latency and jitter delays. The second-tier data centers are located within major metro areas and are used as central points for control and user plane aggregation. They could be 100s of km away from the cell sites. These can be used for internet peering and are usually connected by redundant MPLS rings.

10.4.1.2 Timing Design When DU is at the DC

Figure 10.6 depicts the end-to-end timing design when DU is at the DC.

The solution is based on G8275.1 Full Timing Support (G.8275.1) in the fronthaul with timing from the FH switch or CSR toward other RUs and DU (C3 configuration). The timing from higher level DC to the DC hosting DU is based on Assisted Partial Timing Support (A-PTS) (G.8275.2). All networking gear should be timing-aware.

In this configuration, the FH Switch/CSR acts as the source for SyncE and PTP Master (T-GM) and also acts as a Telecom Boundary Clock (T-BC). At least two sites (three recommended) in a cluster for redundancy are needed. BMCA optimizes the time distribution with multiple

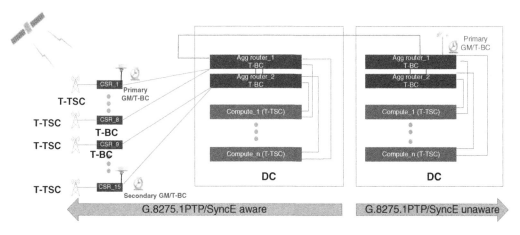

Figure 10.6 Timing distribution when DU is deployed at the DC.

PRTC/T-GMs. The RU and DU act as Telecom Slave Clock (T-TSC) (G.8273.2). Class B is the minimum requirement but it is recommended to have a switch class performance of Class C or higher with a large holdover time.

For a C-RAN deployment, the cell site router drives the timing toward the Top of the Rack (ToR) switch, DU, and other cell site routers. The RU and DU will act as T-TSC and any routers/switches not acting as T-GM should be configured as T-TBC. As basic requirements, T-GM is required to have a built-in GPS receiver for timing that supports PTP distribution compliant to G.8272, and should support IEEE 1588 and SyncE. The system should support T-GM using G8275.1 profile. When GPS is lost, T-GM converts to T-BC and supports ITU-T G.8273.2.

Note: The operator should have a system that is capable of supporting the fiber asymmetry compensation.

10.4.1.3   DU Deployed at Cell Site

In this deployment, the DU is at the cell site. This is still considered O-RAN 7-2x split architecture with LLS. There are two options for the type of compute deployed at the cell site:

- Compute in a Cabinet, e.g. at the base of the tower, not ruggedized but can use fans and support higher temperatures and humidity.
- Ruggedized DU Compute is a fan-less, IP65-compliant unit that does not require a cabinet to maintain environmental conditions. This ruggedized solution can be deployed on the tower itself like a Radio.

Figure 10.7 depicts the end-to-end timing design when DU is at the cell site.

The GPS should connect to an NIC card or CSR in this case and provide timing. The DU can act as a PTP master if GPS connects to the NIC card. The DU can act as a PTP slave if a CSR has PTP GM functionality and can be used to provide timing to both the DU and the RU. To provide redundancy, one has to add the infrastructure as standby.

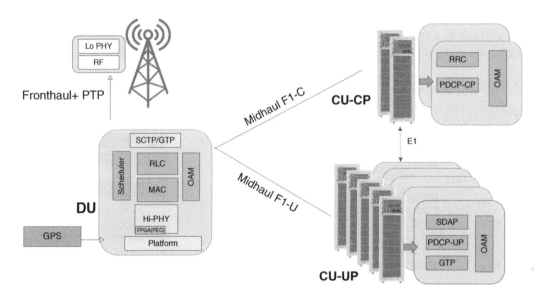

Figure 10.7   DU deployment at site.

10.4.2 CU Deployment

As described earlier, the CU can be deployed in the same or different data center, and connects to the core network. In the case of 5G, UPF can also be deployed in the same location for faster service access. CU can be further split logically or physically into CU-CP and CU-UP. The number of DUs connecting to CU-CP and CU-UP can be different. CU-CP dimensioning is dependent on a number of cells and the number of users. On the other hand, CU-UP dimensioning is based on capacity, i.e. throughput in Gbps and the number of bearers. The placement of the CU components is also dependent on the services being offered, i.e. CU-UP for eMBB and URLLC cases can be located in separate data centers and connected to separate UPFs. A CU-CP is connected to N CU-UPs ($N \geq 1$) and the scaling and LCM of each function can be done independently.

Further, multiple CU functions can be deployed on the same physical node. The CU-UP function usually will have requirements like DPDK for faster data packet processing, SR-IOV and Multus too. It is important for an Operator to check the compatibility of the drivers and the underlying infrastructure for optimal performance.

10.4.3 Radio Unit Instantiation

When an RU comes up online with no pre-existing configuration, it starts scanning for DHCP at every VLAN ID. This results in very long configuration times. In order to reduce RU automation configuration time, it is good practice to plan ahead with a fixed VLAN ID range for both M-Plane and CU-Plane, and with a scan timer. Agreement between the DU and RU vendor is required and both should follow the VLAN tag rules and the DU should tag the VLAN for M-Plane and CU-Plane traffic toward RUs as well. The VLAN ID range should be big enough for the same RU to be connected to multiple DUs, basically allowing mix and match. For example, if there are two DU vendors, four VLAN IDs are needed. Each DU vendor will be assigned a unique M-Plane VLAN ID and a unique CU-Plane ID, and ideally, they should be sequential. So technically, when the RU powers up, it will scan this range only. At the same time, to maintain flexibility and to make it future-proof, the scan range should be bigger to accommodate more DU vendors, beyond just two, maybe a range of 10 VLAN IDs with a scan time of a few seconds (e.g. five seconds) is recommended practice.

The RU establishes a management connection with the Element Management System (EMS) via a specific VLAN. The out-of-box RU shall perform a VLAN scanning procedure to establish successful management connections. The RU shall wait for the DHCP response on each VLAN based on the VLAN Scan Timer before incrementing the VLAN ID. If the RU is configured to scan a VLAN ID in the range of X to Y, it starts with X, X + 1, X + 2 . . . Y VLAN ID to establish a connection.

Upon successful connection establishment with the EMS, the RU shall store the VLAN in its persistent memory. Once the RU has VLAN stored in a persistent memory, it will never use a VLAN scan unless it cannot establish a successful management connection. If the RU requires reboot after a successful connection, the RU shall always use the persistent memory VLAN value to establish a management connection. If the RU is unable to establish the management connection with the VLAN from persistent memory, the RU shall start scanning the complete VLAN list.

Another consideration is IPv4 or IPv6, as part of M-plane configuration, a selection for either IPv4, IPv6, or dual-stack has to be made, and should align between the DU and RU vendors.

10.4.4   Radio Unit Management

The O-RU terminates the Open Fronthaul Interface (i.e. LLS interface) as well as Low-PHY functions of the radio interface toward the UE. The O-RU also terminates the Open Fronthaul M-Plane Interface toward the O-RAN DU (O-DU) and Service Management and Orchestration (SMO) system. There are two models for the Management Architecture of O-RU defined by the O-RAN.WG1.O-RAN-Architecture-Description-v06.00. O-RAN Architecture Description: Hierarchical Management Model, and Hybrid Management Model. Only one Management Mode is active between an O-RU and O-DU combination that has to be selected in advance. If an Operator has multiple vendors in the ecosystem, both models can be supported but not recommended as it does have a direct impact on how the RU is provisioned and managed, integration, and zero touch processes.

10.4.4.1   Hierarchical Management Architecture Model

In this configuration (Figure 10.8), an O-RU is managed by one or more O-DUs. In current specifications, an O-RU is managed by a single O-DU but there is ongoing work in O-RAN for multiple O-DUs to connect to a single O-RU. These O-DUs terminate the monitoring/control of a subordinate O-RU, and are also responsible for managing and configuring O-RU via NETCONF. This does have an operational advantage in that an O-RU can be provisioned without connectivity to a back-end EMS or a Network Management System (NMS), which sometimes may not be available.

10.4.4.2   Hybrid Management Architecture Model

In this configuration (Figure 10.9), an O-RU is managed by one or more EMSs in addition to O-DUs. An advantage of this model is that EMSs can monitor/control other network devices in addition to O-RUs, enabling uniform maintenance, monitoring, and control of all equipment.

In either architectural model, management functions can be limited for each Network Configuration Protocol (NETCONF) client managing an O-RU making for flexible operation. For example, operations can be divided into a NETCONF client performing SW management and a NETCONF client performing fault management.

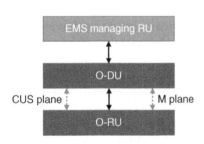

Figure 10.8   Hierarchical RU management.

10.4.5   Network Management

RAN is managed by an EMS that is usually placed centrally in the network and is capable of managing several 100s or 1000s of nodes. In addition to the EMS, they also need DNS and DHCP functions. It is the best practice to avoid configuring IP addresses of each function, one should use Fully Qualified Domain Names (FQDNs) to discover each other. The management architecture should be very simple with minimal components that can scale horizontally as needed. An overly tiered management system is not good and should be avoided. The IP addresses of the RUs are dynamically assigned during the RU bootup time using M-Plane O-RAN-specified procedures and the DHCP servers. The IP address of other

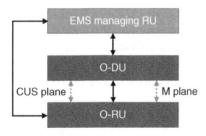

Figure 10.9   Hybrid RU management.

nodes can be assigned statically during the deployment of infrastructure. O-RAN specifies the VLANs, Class of Service (CoS), and DSCP markings for the traffic on the fronthaul. Similarly, based on the deployment scenario and the IP network requirements, VLAN should be configured on the midhaul IP traffic to/from RAN network nodes to segregate different types of traffic on the network. Along with VLAN, CoS and DSCP marking can be configured. Similarly, IP traffic to the core network on the backhaul can be segregated.

Data traffic on the fronthaul between the RU and the DU for C and U Planes uses layer 2 transport using the MAC addresses of RU and DU and a common VLAN. M-Plane traffic on RU is IP layer 3 and uses a separate VLAN for OAM traffic to the EMS and the DU. S-Plane on fronthaul can optionally use VLAN tags. IP subnets and the size of the subnets would depend on the deployment scenario and IP address allocation of IP addresses to RUs.

10.4.6 Public Cloud Provider Overview

This section provides high-level information on the implementation of network components within the public cloud.

10.4.6.1 Native Services

In the public cloud context, the infrastructure is provided by a Public Cloud Provider (PCP), e.g. in the case of Amazon Web Services (AWS), it is provided in the form of Elastic Compute Cloud (EC2) instances. PCPs have a wide flexibility of compute types in terms of networking, processing, and memory sizes that can be used by the operator to right-size the infrastructure. PCPs have several data centers distributed across the country and are well connected with mesh networking. They are designed to be highly available with redundant environmental systems. But the biggest advantage is the availability of consistent infrastructure on demand. Instead of investing in buying the latest compute types every couple of years, the operator can benefit from silicon evolution and can not only experiment and make decisions that suit the current demands but can also quickly adapt to the future ones. PCPs such as AWS are also offering services like AWS Outpost, fully managed and monitored infrastructure that can be deployed on-premises at enterprises or edge locations. The operator also has an option to use the PCP Platform with curated services like AWS Elastic Kubernetes Service (EKS), Google Cloud Platform (GCP), Google Kubernetes Engine (GKE), etc., for running the applications.

Another big advantage is the ability to use native PCP services like Load Balancer, DNS, Container Services, Repository, etc., that are optimized to work with each other out of the box. Similar to infrastructure, these services can be used on demand and are ready to be integrated.

10.4.6.2 CD Pipeline

CI/CD refers to continuous integration and continuous deployment. CI/CD seeks to narrow the gap between development and operation activities and teams by enabling automation in building, testing, and deployment of applications.

For quicker deployment of the virtualized application, Operator can run CI/CD pipeline with infrastructure and platform automation, as part of the zero touch provisioning and LCM. In order for efficient CD pipelines, the vendor has to create a complete package for deployment including pre- and postchecks, health audits, all affinity and anti-affinity, and network requirements. Ideally, CD pipelines only have a point to the repository and the pipelines go through the process of installation, upgrades, etc. Commonly used tools like Ansible, Terraform, Jenkins, etc., can be used for this.

The CD pipelines are a little more complicated if they are running on both public and private clouds and one of them has a dependency or must be executed sequentially. For example, if the CU was to be deployed on public cloud and DU on private cloud, the steps have to be completed for CU instantiation followed by DU. Proper overlay networking needs to be done for DU to discover CU in the public cloud.

One can also add testing as part of the pipeline, specific functional test cases can be run as part of the process. Similarly, CD pipelines can also be run for upgrading the application itself. However, due to the real-time nature of RAN applications like vDU, in-service upgrades are not possible. There is a downtime experienced during the upgrade, alternatively, another instance of vDU can be spun up and traffic can be hard cutover with minimal impact. This obviously requires additional compute which can add to the cost of the network. Other microservices can probably be upgraded without actual service impact.

NF images can be placed in the repository, e.g. AWS Elastic Container Registry (ECR) for the operator to pull them, and perform security scans, and integrity checks before deployment. Access control should be maintained for such repositories for security and management. Whenever a vendor places a new image of the NF, the operator is automatically notified and the workflow engine kicks for seamless deployment. Images can be pulled or pushed depending on how the accounts are set up.

10.4.6.3 Cluster Creation and Management

In the case of Kubernetes, all applications will be placed on worker nodes that are managed by the PCP. In PCP deployment, the operator only has to design the cluster and its instantiation and management are done by PCP itself seamlessly. PCP is responsible for node management and its LCM.

10.4.6.4 Transport Design

For hybrid cloud design, network planning has different considerations. Public Clouds such as AWS and GCP provide several ways to connect. However, for CU-DU connectivity, direct peering services like AWS Direct Connect or Google Interconnect are the best methods for connectivity. Alternative connectivity like VPN will not work due to midhaul latency requirements. Similarly, running DU within the public cloud itself will not be feasible due to latency and jitter requirements of the fronthaul. PCPs offer connectivity on per port and bandwidth basis, this is important to consider because if each vDU connects directly to a PCP direct peering point, there is a cost associated with such connectivity that should be considered but also the peering locations may not have enough ports to handle 1000s of vDUs. It is ideal for an operator to have aggregate vDU traffic at an EDC and send them over to a single port to PCP. At the EDC, all transport circuits within a region can be aggregated by the Service Provider and transported to a remote cloud ingress location over a long-haul provider.

10.4.7 Life Cycle Management of NFs

Virtualized applications like DU and CU are instantiated, upgraded, updated, and terminated via the Life Cycle Manager. The Manager is integrated with CD pipelines that have other managers for infrastructure and platform. The Manager itself is integrated with an Orchestrator that initiates application management based on specific operator policies. The Manager can also configure the application using APIs or Kubernetes operators. A single Manager can also combine application, virtualization, and platform as part of the operation.

The Manager usually has a network service design template and is capable of service chaining based on domain knowledge. The applications are initially on-boarded manually followed by zero touch deployment using workflow automation engines like Ansible. Applications are accompanied by scripts with Cloud Service Archive (CSAR) packages and with the knowledge of topology, the applications are installed on the right nodes.

10.4.8 Network Monitoring and Observability

If an application is built with cloud-native principles, Cloud-Native Computing Foundation (CNCF) offers many open-source system monitoring and alerting tools. These tools are easily pluggable to the applications, i.e. the plugins are widely available. These are communities supported with a very large developer base, and they are lightweight and utilize very little resources. Network observability and the ability to collect performance and faults are critical for RIC operation.

10.4.8.1 Prometheus
Prometheus is widely used for monitoring performance metrics (PM). A Prometheus Agent will be installed for each NF and with a central aggregation engine to collect.

10.4.8.2 Jaeger
Jaeger is another open-source, end-to-end distributed tracing system. Jaeger allows monitoring and troubleshooting of transactions in microservices-based distributed systems. However, please note that Jaeger does not allow running a trace per specific subscriber. It is a "capture-all-or-nothing" model.

Jaeger key features are:

- Distributed transaction monitoring
- Performance/latency optimization
- Root cause analysis

10.4.8.3 Fluentd and Fluentbit
These are lightweight JSON-based Log management tools. It is used for collecting, filtering, buffering, and outputting logs across multiple sources and destinations. They are event-driven and can be used not only for application monitoring but can also be extended for infrastructure monitoring such as CPU, Memory, Storage, and Network usage.

10.4.8.4 Probing
Lightweight probes can be used to monitor customer experience and to run traces to evaluate network performance from end-to-end perspective. These are over-the-top agents can measure and emulate QoE at each subscriber level. The collected data from all the agents are usually collected in a centralized repository for post-processing and filtering, and the data can be transported to data pools or a data lake either as log files or streamed using Kafka or equivalent.

10.4.9 Network Inventory

All network information should be stored in a highly available federated inventory. What cannot be observed cannot be optimized. An operator needs to store an inventory of all virtual and physical functions with a topology view that can be exposed and shared as an API. This information can be used by the Assurance and Orchestration function for optimizing and operating the network.

10.4.10   Building the Right Team

O-RAN is fairly new and the disaggregation and open approach is novel to Telco. It is very important to build the right team to design, deploy, configure, and operate an Open RAN solution. Due to its novelty, it is hard to find the right talent with ready experience. There is no simple method here. Due to the disaggregated nature of the solution, one must have the right resource mix in the team that can work on Cloud, RAN, Platform, Observability, etc., and no longer be limited to just 3GPP technologists. As the network is a reflection of the organization itself, to build a disaggregated, open, and flexible network, the team should work in Devops mode, each task and domain with clear integration and responsibilities. Further, due to the varied nature of the required talent, it may not be possible or wise to build the team in one location. The team can be spread out and work remotely with the right collaboration tool sets. This also eases the burden of finding the right talent in the right place.

In addition to Technology and Engineering acumen, a very deep integration and program management expertise is needed to glue different vendors and groups together. In O-RAN, the operator is also the Prime System Integrator (PSI). The operator has to bring different network components together and stitch them for a working solution. The integration is challenging and time-consuming. There is a significant effort within the O-RAN community to enable interoperability via PlugFests (O-RAN PlugFests) but that is only the beginning. There is a big gap between demonstrating technology and operationalizing it for commercial deployments.

10.5   Conclusion

Building an O-RAN-based network is like a merger of traditional RAN with cloud-based technologies. There are no appliance-based systems anymore, rather everything is running on general compute and requires a deep understanding of network virtualization with a wide variety of deployment options. Integration and observability of O-RAN systems require a different level of understanding and deep expertise.

This chapter provided details on key considerations to deploy O-RAN based networks starting with different network components, virtualization, and cloud-native implementation of the NF to enable network operators to design and build a resilient and reliable network.

Bibliography

3GPP TR 38.806 (2017). *Study of Separation of NR Control Plane (CP) and User Plane (UP) for Split Option 2 (Release 15)*.

3GPP TS 38.101-1 (2022). *NR; User Equipment (UE) Radio Transmission and Reception; Part 1: Range 1 Standalone (Release 17)*.

3GPP TS 38.104 (2022). *NR; Base Station (BS) Radio Transmission and Reception (Release 17)*.

3GPP TS 38.300 (2022). *NR; NR and NG-RAN Overall Description; Stage-2 (Release 17)*.

3GPP TS 38.401 (2022). *NG-RAN; Architecture Description (Release 17)*.

Amplifiers. *Doherty and GaN Amplifier Efficiency Oct 6 2017*. https://www.mwrf.com/materials/article/21848699/doherty-amplifier-combines-high-power-and-efficiency.

BASTA Standard Recommendation on Standards for Passive Base Station Antennas BASTA standard v12 Dec 23 2021. https://www.ngmn.org/wp-content/uploads/BASTA-Passive-Antennas-WP12.0.pdf.

G.8273.2. *Timing Characteristics of Telecom Boundary Clocks and Telecom Time Slave Clocks for Use with Full Timing Support from the Network.*

G.8275.1. *Precision Time Protocol Telecom Profile for Phase/Time Synchronization with Full Timing Support from the Network.*

G.8275.2. *Precision Time Protocol Telecom Profile for Time/Phase Synchronization with Partial Timing Support from the Network.*

NEBS Requirements. https://telecom-info.njdepot.ericsson.net/site-cgi/ido/docs.cgi?ID=SEARCH&DOCUMENT=GR-63.

NETCONF. *The Network Configuration Protocol (NETCONF) Provides Mechanisms to Install, Manipulate, and Delete the Configuration of Network Devices.*

O-RAN PlugFests support the ecosystem players in testing and integration of their implementations, ensuring the openness and interoperability of O-RAN solutions from different providers.

O-RAN.WG1.O-RAN-Architecture-Description-v06.00 (2022). *O-RAN Architecture Description.*

O-RAN.WG2.Non-RT-RIC-ARCH-TR-v01.01 (2021). *Non-RT RIC Functional Architecture Technical Report.*

O-RAN.WG2.Non-RT-RIC-ARCH-TS-v02.01 (2022). *O-RAN Working Group 2 (Non-RT RIC and A1 Interface WG) Non-RT RIC Architecture.*

O-RAN.WG3.E2SM-RC-v01.03 (2022). *O-RAN Work Group 3 Near-Real-time RAN Intelligent Controller and E2 Interface.*

O-RAN.WG3.RICARCH-v02.01 (2022). *Functional Architecture of O-RAN Near-Real-Time RAN Intelligent Controller (Near-RT RIC).*

O-RAN-WG4.MP.0-v09.00 (2022). *O-RAN Fronthaul Working Group Management Plane Specification.*

RICAPP. *RAN Intelligent Controller Applications.* https://docs.o-ran-sc.org/en/dawn/projects.html#ran-intelligent-controller-applications-ricapp.

RoE. *Standard for Radio Over Ethernet (RoE) Encapsulations and Mappings.* https://www.ieee1904.org/3/tf3_home.shtml.

YANG is a Data Modeling Language Used to Model Configuration and State Data Manipulated by the NETCONF.

11

Open RAN Test and Integration

Ian Wong, Ph.D.

VIAVI Solutions, Austin, TX, USA

11.1 Introduction

There is a lot to like about Open Radio Access Networks (O-RANs) (see Figure 11.1). Founded on principles of openness, disaggregation, virtualization, and intelligence, this industry movement stands to be a tectonic shift in how radio access networks are developed, tested, deployed, and operated, affecting all aspects of the network life cycle. It frees mobile operators and private network operators to choose hardware and software for the various network functions from multiple vendors rather than being locked into a single supplier's ecosystem. That can reduce capex and opex while providing operational flexibility and supplier diversity, hence increasing competition and motivation for increased innovation.

But there is also a lot to fear, starting with complexity. By allowing various network functions and their component hardware and software pieces to be potentially supplied by different vendors, the issue of interoperability and performance comes to the forefront. An operator cannot afford degradation in service quality for their networks. Hence, open RAN implementations need to satisfy the same robustness and reliability requirements as traditional single-vendor RANs. Also in contrast to single-vendor RANs, the absence of a "single neck to choke" when network issues arise means either the operators themselves should shoulder more of the burden of integration and test or ecosystem partners like system integrators (SIs) will be called upon to play a bigger role in ensuring a smooth network deployment and rollout. Furthermore, by moving more and more components to be virtualized and cloud-native, and relying on AI/ML technologies for increased network intelligence and automation, the complexity is further compounded by the highly dynamic nature of the continuous development of new software features.

These concerns and challenges highlight the importance of testing early and often, from lab validation to field verification, and finally onto network assurance and optimization. This chapter delves into the challenges of testing Open RAN and how to address them. It then outlines the key initiatives undertaken by the O-RAN ALLIANCE to support the ecosystem in overcoming these challenges.

Open RAN: The Definitive Guide, First Edition. Edited by Ian C. Wong, Aditya Chopra, Sridhar Rajagopal, and Rittwik Jana.

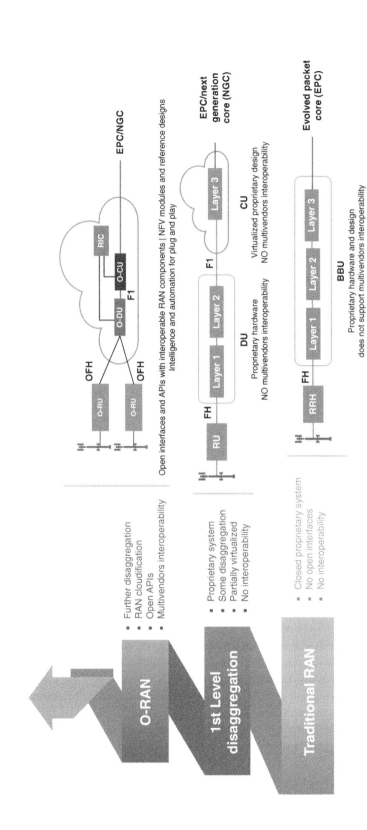

Figure 11.1 The O-RAN evolution fundamentally changes the mobile network architecture. *Source:* VIAVI Solutions, used with permission.

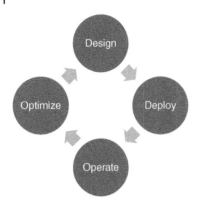

Figure 11.2 Basic network life cycle.

11.2 Testing Across the Network Life Cycle

Figure 11.2 shows a basic network life cycle, comprising the key steps of:

1) Design
2) Deploy
3) Operate
4) Optimize

The "design" phase involves understanding the business requirements of the network and designing the key components to fulfill those requirements. The design phase, in the macro sense, involves network operators planning their networks, and equipment manufacturers doing the necessary research and development (R&D) to design, develop, and manufacture the equipment needed to fulfill the requirements of these networks. In the era of 5G and Open RAN, network operators not only include the typical mobile network operator (MNO) or service provider (SP), but also the private network operator which could be an Enterprise or a University campus, or the SI tasked with integrating, testing, and, in some cases, operating and optimizing these networks. Similarly, "equipment manufacturers" in the Open RAN sense not only include the traditional fully integrated equipment vendors, but also the white-box hardware vendors that supply general-purpose chips and servers, the software vendors supplying the various network functions that are virtualized or cloud-native, and even the hyperscale cloud providers (HCPs) and management and orchestration software and SPs that provide the infrastructure to run and operate the network.

This complex ecosystem of operators and vendors needs to come together to form an interoperable, robust, reliable, yet open network. This necessitates rigorous interface specifications, comprehensive test specifications around these interfaces, and laboratory test infrastructure that can validate the implementations continuously and comprehensively, while having a path toward continuous monitoring and optimization post deployment.

The "deploy" phase, as the name suggests, involves the deployment of the carefully designed network in the field. In the case of Open RAN, careful verification that all the network components are appropriately installed and turned up, and the desired RF propagation characteristics are indeed achieved in the field. Given the complexity of a multivendor network, it is imperative that every interface is carefully monitored for potential anomalies, which includes the entire transport network and their timing and synchronization.

Finally, in the "Operate/Optimize" phase, the network operator will need to ensure the correct network parameters are indeed being used across the network, closely monitor the performance, and optimize the network to ensure adequate quality of service to its end users while ensuring energy efficiency and reliability. In the case of Open RAN, the ability to use open interfaces for RAN intelligence monitoring and control opens up the possibilities for fully automated networks that require minimal human intervention.

All this comes back to the design phase for continuous improvement of the network, and where agile software development practices, i.e. continuous integration/continuous deployment (CI/CD), are practiced so newer and better services can be delivered to end users while preserving robustness and reliability of the network.

11.3   O-RAN ALLIANCE Test and Integration Activities

In order to support the Open RAN ecosystem with its test and integration challenges, the O-RAN ALLIANCE, embarks on several key initiatives:

1) Development of Test Specifications
2) Development of processes for Certification and Badging of vendor equipment
3) Support establishment of OTICs
4) Organizing the O-RAN ALLIANCE Global PlugFests

11.3.1   Test Specifications

The O-RAN ALLIANCE develops normative technical specifications around key interfaces, procedures, and requirements to foster interoperability among the different network functions that comprise an Open Radio Access Network. Figure 11.3 shows a functional diagram for a complete 3GPP-based open radio access network implementation, including the main O-RAN and 3GPP interfaces. As the O-RAN ALLIANCE is focused on being complementary to 3GPP technical and test specifications, it is only focused on developing technical and test specifications on O-RAN standardized elements and does not repeat nor conflict with existing 3GPP specifications.

Traversing Figure 11.3 from left to right, the user equipment (UE) shown on the left are connected to the network via the *Uu* air interface, which is standardized in 3GPP, to the open radio units (O-RUs) via radio frequency (RF) propagation. The figure shows both an LTE path and a 5G path for illustration purposes. The O-RUs are connected via the open fronthaul (OFH) control, user, synchronization, and management (CUSM) planes to the open distributed units (O-DU), and optionally through a fronthaul multiplexer (FHM). More details on the OFH can be found in

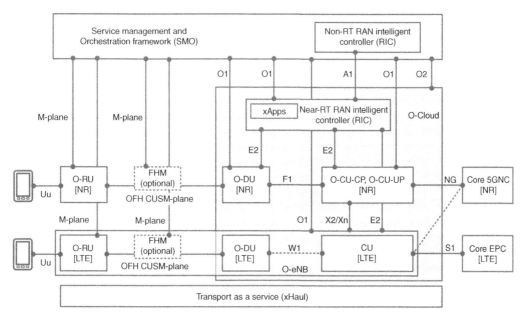

Figure 11.3   Illustrative open RAN functional diagram. *Source:* Adapted from O-RAN ALLIANCE / https://www.o-ran.org/blog/o-ran-global-plugfest-2021-demonstrates-stronger-ecosystem-and-maturing-solutions.

Chapter 6. The O-DU is then further connected to the open centralized unit (O-CU) control plane (O-CU-CP) and user plane (O-CU-UP), via the F1 interface (W1 for LTE), which is standardized in 3GPP RAN3, but where the O-RAN ALLIANCE Working Group 5 develops interoperability profiles. Additionally, the X2/Xn interfaces for inter-eNB/gNB communication are also standardized in 3GPP RAN3 with profiles developed in O-RAN ALLIANCE Working Group 5.

The O-CU then subsequently connects to the 5G Core (enhanced packet core [EPC] for LTE) network via the NG-C/U (S1 for LTE) interfaces which are standardized in 3GPP.

Now traversing Figure 11.3 from top to bottom, the service management and orchestration (SMO) framework uses the O1 interface as the main network management interface to all the network functions of the entire network (except for the O-RU, which is managed via the OFH M-plane), which is standardized in Working Group 10 in O-RAN ALLIANCE. A key aspect of the O-RAN architecture is the RAN intelligent controller (RIC) functionality, where a non-real-time RIC within the SMO connects to the near-real-time RIC via the A1 interface, and subsequently the near-RT RIC to the O-DU/O-CU (collectively E2 nodes) via the E2 interface. All the DU/CU/RIC network functions can be run in cloud-based infrastructure, which is managed by the SMO through the O2 interface standardized in O-RAN Working Group 6. Finally, all these network functions need a reliable transport network comprising of front-, mid-, and back-haul, collectively known as Xhaul, whose requirements are defined in O-RAN Working Group 9. Table 11.1 summarizes the key interfaces outlined above from both 3GPP and O-RAN ALLIANCE and their corresponding technical and test specifications for easy reference.

For the rest of this section, we will focus on the test specifications from the O-RAN ALLIANCE for the various interfaces/subsystems described earlier, which are categorized into three main classes:

1) Conformance Test Specifications
2) Interoperability Test Specifications
3) End-to-end Test Specifications

These are described in the sections as follows.

11.3.2 Conformance Test Specifications

Conformance test specifications focus on verifying the conformance of a single device under test (DUT) to technical specifications. In O-RAN, normative technical specifications cover not only O-RAN-specified interfaces, e.g. OFH, E2, A1, etc., but also systems, e.g. Xhaul transport and security. Conformance test specifications are meant to verify that implementations are compliant with these normative technical specifications. In the case of interface specifications, these involve at least two entities that are communicating over the interface, the conformance assessment will need to be performed on each of the different entities, and the test configuration for each entity will also be different.

Conformance test specifications outline the test setup, entrance and exit criteria, and pass/fail criteria for each specific functionality that is being tested. The tests follow the principle of not putting the DUT into any special test modes, i.e. the DUT should be configured in operational mode, so as not to burden implementations to develop specific software/firmware only for the purposes of testing. It is critical that a clear pass-fail criteria is set up for each test that attests the correct implementation of each feature according to the specification. Furthermore, since the tests are primarily for baseline conditions with the most basic parameters, it attempts to ensure interoperability assuming the DUTs are configured with these basic parameters. Since actual network deployments will likely deviate from these baseline parameters, tests beyond conformance are necessary.

Table 11.1 Interfaces in 3GPP-based open radio access network with corresponding technical and test specifications.

Interface	Endpoints	SDO responsible	Technical specifications	Test specifications	Description
Uu (RF)	UE ↔ O-RU	3GPP RAN 1, 2, 4	3GPP TS 38.2xx/38.3xx 36.2xx/36.3xx (LTE)	3GPP TS 38.141-1/2 36.141 (LTE)	RF air interface between UE and RAN through the O-RU
Open fronthaul (OFH) CUSM-planes	O-RU ↔ O-DU	O-RAN WG4	O-RAN.WG4.CUS O-RAN.WG4.MP	O-RAN.WG4.CONF O-RAN.WG4.IOT	Open fronthaul interface between O-RU and O-DU
O1	SMO ↔ managed NFs	O-RAN WG10	O-RAN.WG10.O1-Interface O-RAN.WG5.O-DU-O1 O-RAN.WG5.O-CU-O1	TBD	Management and measurement reporting interface between SMO and all managed NFs
E2	Near-RT RIC ↔ E2 Node	O-RAN WG3	O-RAN.WG3.E2xxx	O-RAN.WG3.E2TS	Interface connecting near-RT RIC and E2 nodes (O-DU, O-CU, O-eNB)
F1-C/U	O-DU ↔ O-CU	3GPP RAN3 O-RAN WG5	3GPP TS 38.47x O-RAN.WG5.C O-RAN.WG5.U	O-RAN.WG5.IOT	3GPP-defined interfaces between gNB-DU and gNB-CU-CP and UP, respectively, and adopted by O-RAN under defined interoperability profiles
X2-C/U	eNB ↔ eNB en-gNB ↔ eNB	3GPP RAN3 O-RAN WG5	36.42x O-RAN.WG5.C O-RAN.WG5.U	O-RAN.WG5.IOT	3GPP-defined interface
Xn-C/U	gNB ↔ gNB ng-eNB ↔ ng-eNB/gNB	3GPP RAN3 O-RAN WG5	38.42x O-RAN.WG5.C O-RAN.WG5.U	O-RAN.WG5.IOT	
A1	Non-RT RIC ↔ Near-RT RIC	O-RAN WG2	O-RAN.WG2.A1xxx	O-RAN.WG2.A1TS	Interface to enable non-RT RIC to provide to near-RT RIC policy-based guidance, ML model management and enrichment information
O2	SMO ↔ O-Cloud	O-RAN WG6	O-RAN.WG6.O2xxx	TBD	
NG-C/U	O-CU-CP/UP ↔ 5GC	3GPP	3GPP TS 38.41x		
S1	LTE CU ↔ EPC	3GPP	3GPP TS 36.41x		
CTI	O-DU ↔ Transport Node	O-RAN WG4	O-RAN.WG4.CTIxxx	TBD	

As of this writing, the O-RAN ALLIANCE has five active conformance test specifications:

1) O-RAN Working Group 2 A1 Interface Test Specification
2) O-RAN Working Group 3 E2 Interface Test Specification
3) O-RAN Working Group 4 Open Fronthaul Conformance Test Specification
4) O-RAN Working Group 9 Xhaul Transport Testing
5) O-RAN Working Group 11 Security Test Specifications

These are outlined in the subsections as follows.

11.3.2.1 A1 Interface Test Specification (O-RAN.WG2.A1TS)

This specification covers the conformance of the non-real-time RAN intelligent controller (non-RT RIC) or the near-real-time RAN intelligent controller (near-RT RIC) to the A1 interface technical specifications. The A1 interface enables the non-RT RIC to provide policy-based guidance, machine learning (ML) model management and enrichment information to the near-RT RIC (see Chapter 5 for more details on the RIC.) The test configurations for the non-RT RIC and near-RT RIC are shown in Figures 11.4 and 11.5.

Non-real-time RIC (Device under test)

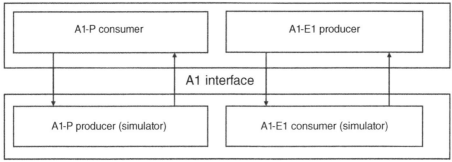

Test simulator

Figure 11.4 Test configuration for A1 conformance of the non-RT RIC. *Source:* O-RAN.WG2.A1TS-v03.00.

Test simulator

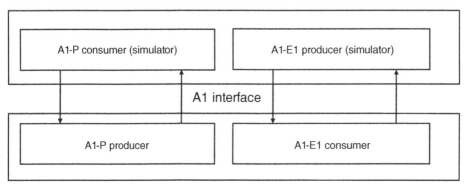

Device under test near-real-time RIC

Figure 11.5 Test configuration for A1 conformance of the near-RT RIC. *Source:* O-RAN.WG2.A1TS-v03.00.

These conformance tests require emulators of the peer functionality for the interfaces under test for the given DUT, and should have capabilities of generating HTTP requests and responses, and flexibility in configuring headers and body for these HTTP requests and responses to enable the creation of various test cases.

11.3.2.2 E2 Interface Test Specification (O-RAN.WG3.E2TS)

This specification covers the conformance of the near-RT RIC or the E2 Nodes (i.e. O-eNB, O-DU, O-CU-CP, and O-CU-UP) to the E2 interface technical specifications. The test setup for E2 conformance for the near-RT RIC and E2 Nodes as the DUT are shown in Figures 11.6 and 11.7, respectively.

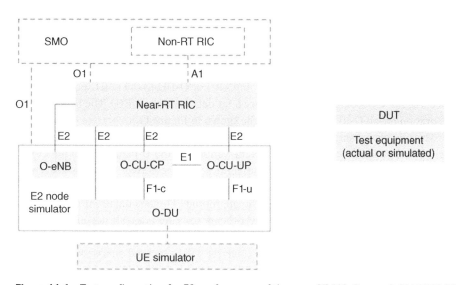

Figure 11.6 Test configuration for E2 conformance of the near-RT RIC. *Source:* O-RAN.WG3.E2TS.v02.00.

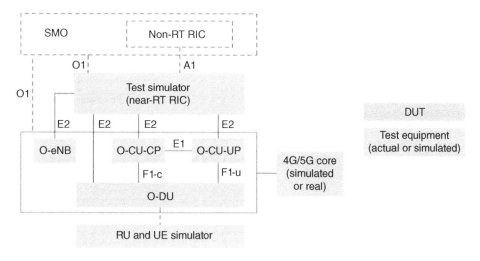

Figure 11.7 Test configuration for E2 conformance of the E2 nodes. *Source:* O-RAN.WG3.E2TS.v02.00.

Although similar in principle to the A1 interface conformance testing, the E2 interface testing is slightly more complex due to the need to emulate various E2 nodes when testing the near-RT RIC, which require to be stateful, realistic emulations, and conversely emulating the near-RT RIC as seen by the various E2 nodes. In addition, the emulators also need to support various RAN configurations as defined in (*O-RAN.WG1.O-RAN-Architecture-Description, Annex A.4*), e.g.

- Standalone O-CU-CP connected to one or more standalone O-CU-UP and one or more standalone O-DU. Each logical node is considered as an E2 Node that presents an E2 interface to the Near-RT RIC.
- Combined O-CU-CP and O-CU-UP connected to one or more standalone O-DU. The combined O-CU-CP/O-CU-UP may present either a common E2 interface or individual E2 interfaces corresponding to the individual O-RAN components.
- Combined O-CU-CP, O-CU-UP, and O-DU. The combined node may present either a common E2 interface or individual E2 interfaces corresponding to the individual O-RAN components.

11.3.2.3 Open Fronthaul Conformance Test Specification (O-RAN.WG4.CONF)

This specification covers the conformance of either an O-RU or O-DU to the OFH specifications, covering all aspects of the CUSM planes (see Chapter 6 for details on the OFH). The baseline CU plane test configuration for an O-RU under test can be shown in Figure 11.8.

The basic procedure involves using an O-DU CUSM-Plane Emulator connected via the fronthaul to the O-RU DUT. The CUSM-Plane Emulator sends the various C- and U-plane data to the O-RU DUT, while the RF signal analyzer on the other end verifies that the correct output is observed. Conversely, the RF signal generator transmits the appropriate RF signals while the CUSM-Plane emulator sends the C-plane commands to observe the appropriate U-plane data sent up from the O-RU DUT. Note that options for RF conducted or over-the-air tests are available depending on the O-RU capabilities, i.e. availability of conducted ports. The S- and M-plane test configurations are quite similar, except that for the S-plane tests, the RF analyzer will be focused on assessing the synchronization capabilities of the O-RU, while in the M-plane tests, only the O-DU emulator is required with some other peripheral circuitry to verify the specific features of an O-RU, e.g. external I/O lines.

When the O-RU is the Device Under Test (DUT), there is Test Equipment, RU ("TER") that is used to connect to the DUT, run the test and evaluate the test result.

The Test Equipment, RU ("TER") provides everything needed to operate the test, including:

- M-Plane commands to collect O-RU capabilities and configure the O-RU

- Synchronization via G.8275.1 (Mandatory for O-RU)

- C-Plane and U-Plane data flow to the O-RU

- Collection of RF energy to determine if the O-RU reacted correctly to the DL C-Plane and U-Plane data

- Radiation of RF energy to be received by the O-RU (note: this could be actually radiated, or conveyed through multiple RF cabled connections)

- Collection of O-RU UL fronthaul data flows and evaluating it for correctness based on the supplied RF energy to the O-RU antennas or RF connectors

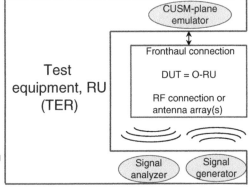

Figure 11.8 Test configuration for open fronthaul conformance of the O-RU. *Source:* O-RAN.WG4. CONF.0-v07.00.

O-DU DUT

Combined O-CU+O-DU DUT

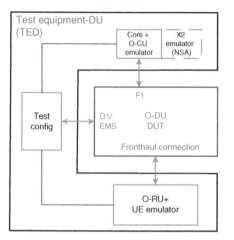
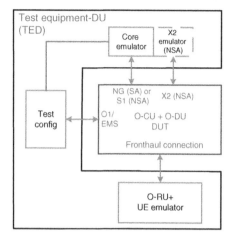

Figure 11.9 Test configuration for open fronthaul conformance of the O-DU. *Source:* O-RAN.WG4. CONF.0-v07.00.

The test configuration for an O-DU is a bit more complex, as shown in Figure 11.9. Since the O-DU supports various interfaces depending on the network configuration, the diagram on the left shows the configuration for a 5G standalone (SA) O-DU, and the right diagram shows the configuration for a 5G nonstandalone configuration (NSA). Furthermore, following the principle of not requiring the DUTs to have special test modes, the test emulation environment requires fairly complete functionality so as to push the O-DU DUT through the various states to support a standard UE connection. This then allows the O-RU + UE emulator to observe the CUSM plane signals from the O-DU DUT to verify the correct functional implementation of the O-DU fronthaul.

11.3.2.4 Xhaul Transport Testing (O-RAN.WG9.XTRP-Test.0)

This specification covers tests for an Open Xhaul (i.e. front/mid/backhaul) transport infrastructure (see Chapter 7 for more details on open transport). It includes cases that test:

1) Xhaul networks from a universal network interface (UNI) interface to another UNI interface.
2) Xhaul network segments that carry traffic for an overlay network.
3) Individual transport network equipment or their components.

Details on transport testing can be found in Chapter 7.

11.3.2.5 Security Test Specifications (O-RAN.SFG.Security-Test-Specifications)

This test specification is focused on validating the proper implementation of security protocol requirements specified in (*O-RAN.SFG.Security-Protocols-Specification*). See Chapter 8 for further details on Open RAN Security.

11.3.3 Interoperability Test Specifications

Interoperability test specifications focus on assessing the interoperability of a pair of DUTs across an interface. One of the pillars of O-RAN is the interoperability of various network functions across standardized interfaces. Thus, having well-established interoperability test specifications is extremely important, and goes beyond conformance testing, which is what other standards development organizations typically specify. Interoperability tests extend beyond the basic functional assessment of an interface implementation using the most basic parameters, toward assessing the performance of the interface under more realistic parameter configurations.

At the time of this writing, there are three active interoperability test specifications:

1) Fronthaul Interoperability Test Specification (*O-RAN.WG4.IOT.0-09.00*)
2) Open F1/W1/E1/X2/Xn Interoperability Test Specification (*O-RAN.WG5.IOT.0-v06.00*)
3) Stack Interoperability Test Specification (*O-RAN.WG8.IOT-v06.00*)

11.3.3.1 Fronthaul Interoperability Test Specification (O-RAN.WG4.IOT.0-09.00)

This specification is focused on assessing the interoperability between an O-DU and O-RU over the OFH interface. The tests cover all CUSM planes of the OFH. Figure 11.10 shows a typical interoperability test configuration, where the system under test (SUT) here are both the O-DU and O-RU and their respective OFH interface connection. The diagram on the left shows a SUT configuration where the O-DU is a standalone implementation with a northbound F1 interface, and is thus bookended by a Core + CU Emulator on the top and a UE emulator at the RF interface of the O-RU. The diagram at the right is for a SUT configuration where the O-CU and O-DU are combined and the northbound interface is directly to a Core emulator.

This specification also includes specific parameter configurations used for the interoperability testing of the OFH. These configurations are termed "interoperability test (IOT) profiles" and are sets of parameters that are carefully selected to encompass a fairly comprehensive coverage of the main features of the interface under more realistic deployment scenarios. Note that these profiles are not a reflection of actual operator deployment scenarios since doing so will be intractable, but are instead representative parameter sets that should provide a good indication, in a broad sense, of the typical operator deployment scenarios. The current version of the spec (*O-RAN.WG4. IOT.0-09.00*) includes IOT profiles for various M-plane scenarios encompassing hierarchical and

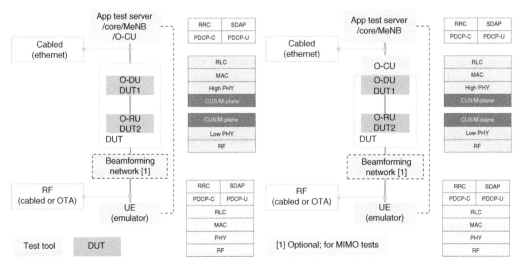

Figure 11.10 Test configuration for open fronthaul interoperability between an O-RU and standalone O-DU (left) or O-DU/CU combined implementation (right). *Source:* O-RAN.WG4.IOT.0-v09.00.

hybrid architectures (*O-RAN-Architecture-Description-v08.00*), and CU-plane scenarios covering 5G NR and 4G LTE TDD and FDD.

11.3.3.2 Open F1/W1/E1/X2/Xn Interoperability Test Specification (O-RAN.WG5.IOT.0)

This specification defines interoperability testing for the radio access network nodes from different vendors connected using the X2, F1, and Xn interfaces implemented in accordance with the NR C-Plane (*O-RAN.WG5.C.1-v09.00*) and U-Plane (*O-RAN.WG5.U.0-v05.00*) profiles. Note that these interfaces are 3GPP interfaces, and O-RAN ALLIANCE Working Group 5 defines profiles for these interfaces to facilitate the interoperability of these network nodes, and this IOT test specification outlines the methodology and procedures for testing.

Figures 11.11–11.13 show the basic diagrams for X2, F1, and Xn interoperability test configurations. As can be seen, the basic SUT configurations are the DUT pairs and the interface that connects them, and the test principles of wrapping around the SUT with emulators that allow the DUTs to function as if in actual deployment in order to ascertain the interoperability of the DUTs over the specific interface.

11.3.3.3 Stack Interoperability Test Specification (O-RAN.WG8.IOT)

In contrast to the prior two interoperability test specifications, which are focused on verifying the interoperability across certain interfaces, this specification is focused on the interoperability between actual O-DU and O-CU implementations, involving much more detailed procedures that involve the internal implementation of the O-DU and O-CU themselves. However, since the O-CU

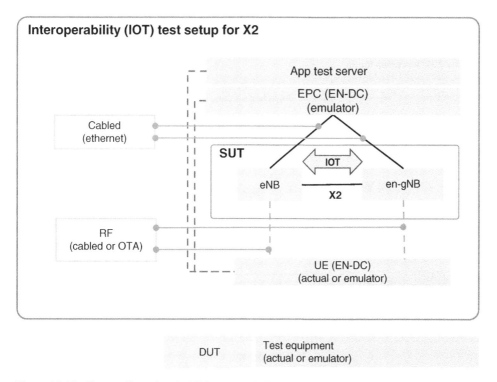

Figure 11.11 Test configuration for X2 interoperability between an eNB and en-gNB. *Source:* O-RAN.WG5. IOT.0-v06.00.

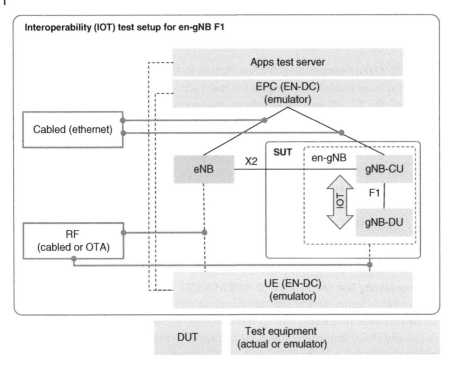

Figure 11.12 Test configuration for F1 interoperability between gNB-CU and gNB-DU. *Source:* O-RAN.WG5. IOT.0-v06.00.

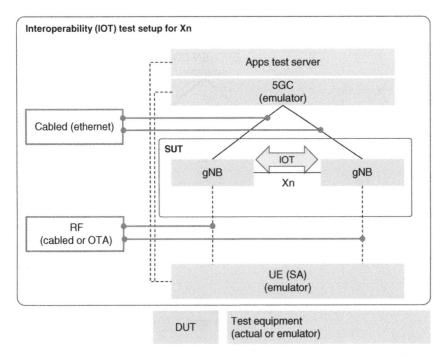

Figure 11.13 Test configuration for Xn interoperability between gNBs. *Source:* O-RAN.WG5.IOT.0-v06.00.

and O-DU stack reference designs being developed by WG8 in O-RAN ALLIANCE are informative and not normative specifications, this test specification is likewise informative in nature. The following interfaces (logical or physical) are covered in IOT test cases:

- RLC-MAC
- L1-L2
- RRC-SDAP
- SDAP-PDCP
- RRC-PDCP
- O1 interface
- E2 interface

11.3.4 End-to-End Test Specifications

The O-RAN end-to-end test specification (*O-RAN.TIFG.E2E-Test.0-v04.00*) is focused on validating the end-to-end system functionality, performance, and key features of the O-RAN system as a black box in both lab and field environments. The internal functionality and architecture of the SUT are out of scope and not being verified explicitly. Figure 11.14 shows the baseline test configuration, which is an O-RAN system book ended by a UE/UE emulator and a Core network/core emulator.

All the components of the O-RAN base station (e.g. O-CU, O-DU, O-RU, and RIC) and their respective interfaces (e.g. OFH and X2) are recommended to have been already node tested against their respective conformance and interoperability test specifications if they are available.

The specification focuses on the following tests:

1) **Functional:** assess the baseline functionality of the O-RAN system, including attaching/detaching and registration/deregistration of a single/multiple UEs, various mobility scenarios including idle and connected mode mobility for intra- and inter-O-DU/O-CU and 5G-LTE mobility scenarios.
2) **Performance:** assess the performance of the O-RAN system, including downlink and uplink peak throughput in various radio conditions for single/multicells and single/multiple UEs, and assess the impact of front/midhaul latency on these throughputs.
3) **Services:** assess the performance of the O-RAN system in the context delivering specific network services, i.e. data service, video streaming, voice, video calling, and services supported using URLLC and mMTC.
4) **Security:** assess the security aspects of the O-RAN system, which include the gNB security assurance specification required by 3GPP SA3 (*3GPP TS 33.511*), denial of service, fuzzing, and exploitation test cases, and resource exhaustion security test cases against underlying cloud infrastructure.

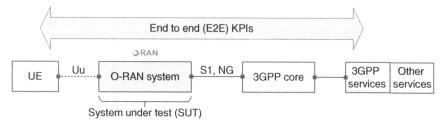

Figure 11.14 Test configuration for an O-RAN system end-to-end test. *Source:* O-RAN.TIFG. E2E-Test.0-v04.00.

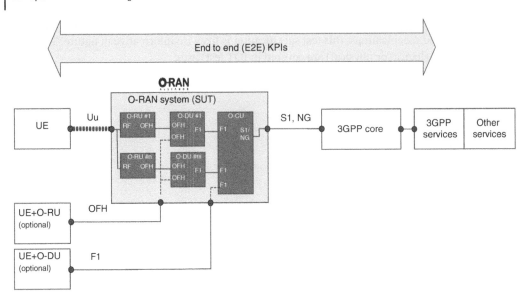

Figure 11.15 Test configuration for an O-RAN system end-to-end load and stress test. *Source:* O-RAN.TIFG. E2E-Test.0-v04.00.

5) **Load and stress:** assess the stability and reliability of the O-RAN system under load and stress conditions, e.g. maximum number of RRC-connected UEs, maximum traffic load, long-hour stability, and multiple cells. These tests are conducted in a lab environment using emulators, as shown in Figure 11.15. In an actual O-RAN system, multiple O-RUs are typically connected to an O-DU, and subsequently, multiple O-DUs connect to an O-CU. In order to truly load an O-CU, for example, a fairly large number of O-DUs and an even larger number of O-RUs need to be loaded up with maximal UE traffic, which makes the test setup overly expensive and cumbersome. In order to alleviate this, the ability of the test system to emulate UE traffic not just over the RF interface to an O-RU, but also simultaneously over the OFH interface to an O-DU(s) and the F1 interface to the O-CU would be highly beneficial.

6) **RIC-enabled use case:** assess the performance of the O-RAN system including the non-RT and near-RT RIC components under specific use case scenarios, e.g. traffic steering. Figure 11.16 shows the test setup for these test cases, where the SUT includes the RIC components in addition to the O-DU and O-CU. The O-RU is not recommended to be part of the SUT since it does not contribute materially to the outcome of these tests, and several O-RUs would be needed to sufficiently extract the performance benefit of the RIC for each of the use cases. The test needs to evaluate UE and system performance for mixed types of service, e.g. voice service, video service, data service, etc.

11.3.5 O-RAN Certification and Badging

In order to further support the Open RAN ecosystem, the O-RAN ALLIANCE has also developed a certification and badging program (*O-RAN.TIFG.Cert-Badge.0-v06.00*) that utilizes the test specifications and allows vendors to show that their equipment is indeed compliant to the O-RAN

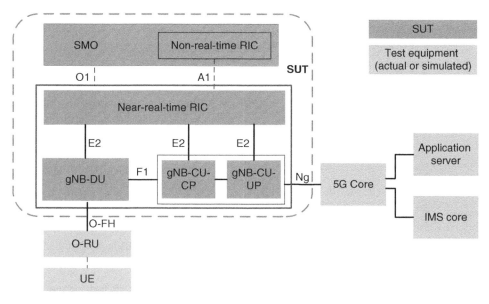

Figure 11.16 Test configuration for an O-RAN system end-to-end RIC-enabled use case test. *Source:* O-RAN.TIFG.E2E-Test.0-v04.00.

technical specifications and interoperable with other vendor's equipment. The three main program components are:

1) **Compliance certification:** A compliance certificate is issued to equipment that have passed all the required tests for the specific interface/feature according to the conformance test specifications. A certificate is issued for the specific interface/feature on a single piece of equipment.
2) **Interoperability badging**: An interoperability badge is issued for the pair of equipment that have passed all the required interoperability tests according to the interoperability test specifications. The specific interoperability test profile(s) passed by the equipment pair for the specific interface under test should be clearly indicated.
3) **End-to-end badging**: An end-to-end badge is issued for the group of equipment that have passed all the required end-to-end tests according to the end-to-end test specification.

Although the testing can be performed by any vendor in any laboratory of their choosing (even their own laboratory) and can claim that they performed the tests according to the procedures, only official OTICs (see Section 11.4.3) can officially issue the O-RAN certificates and badges.

A summary of the certification and badging structure and how it relates to the O-RAN technical and test specifications is shown in Figure 11.17. As of this writing, not all the test specifications have an official certification and badging program associated with them. Table 11.2 shows the available certificates and badges as of this writing (*O-RAN.TIFG.Cert-Badge.0-v06.00*).

11.3.6 Open Test and Integration Centers

One of the biggest advantages of the Open RAN movement is the fostering of a diversity of equipment vendors (including hardware, software, and services) large and small. That being said, there are inherent challenges with testing and integration esp. for smaller vendors, e.g.

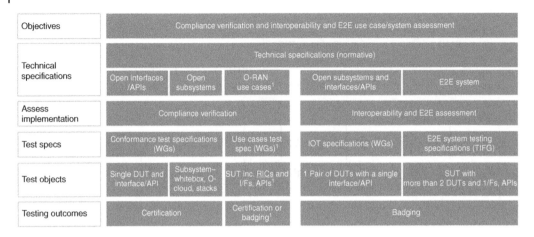

Figure 11.17 O-RAN certification and badging structure. *Source:* O-RAN.TIFG.Cert-Badge.0-v06.00.

Table 11.2 Available certificates and badges and their corresponding certification and badging spec section and test specification section.

Category	DUT/SUT	Certification and badging specification section (O-RAN.TIFG. Cert-Badge.0-v06.00)	Test specification section
Certificates	Open fronthaul for O-RU	Section 4.4	(*O-RAN.WG4.CONF.0-v07.00*) CU plane: section 3.2 S-plane: section 3.3.2–3.3.4 M-plane: section 3.1
	Open fronthaul for O-DU	Section 4.5	(*O-RAN.WG4.CONF.0-v07.00*) CU plane: section 3.4 S-plane: section 3.3.5–3.3.8 M-plane: section 3.1
Interoperability badges	Open fronthaul between O-DU and O-RU	Section 5.2.1	(*O-RAN.WG4.IOT.0-v09.00*)
	X2 interface between eNB and en-gNB	Section 5.2.2	(*O-RAN.WG5.IOT.0-v06.00*) section 2.2
	F1 interface between gNB-CU and gNB-DU	Section 5.2.3	(*O-RAN.WG5.IOT.0-v06.00*) section 2.3
	Xn interface between 2 gNBs	Section 5.2.4	(*O-RAN.WG5.IOT.0-v06.00*) section 2.4
End-to-end badge	Complete O-RAN base station	Chapter 6	(*O-RAN.TIFG. E2E-Test.0-v04.00*)

availability of test equipment and expertise, availability of partners to do interoperability and end-to-end testing, access to operator customers, etc. It is with this in mind that the O-RAN ALLIANCE have developed the concept of an Open Test and Integration Center, or OTIC for short. OTICs are test laboratories hosted by vendor-independent (i.e. operators, test services companies, and academic research labs) O-RAN member companies that can support a variety of services, including but not limited to:

1) Hosting O-RAN PlugFests (see Section 11.4.4)
2) Perform Certification and Badging Services
3) Support the ecosystem through workshops and tutorials

These laboratories are open to all to avail of their services and are envisioned to be geographically diverse. Every OTIC is a separate entity from the O-RAN ALLIANCE but is bound to the rules and regulations through an OTIC-hosting agreement (*O-RAN.TIFG.CGofOTIC.0-v04.00*). As of this writing, there are seven officially approved OTICs all around the world:

1) European OTIC in Berlin hosted by Deutsch Telecom
2) European OTIC in Torino hosted by Telecom Italia
3) European OTIC in Madrid hosted by Telefonica
4) European OTIC in Paris hosted by Orange Telecom
5) Auray OTIC and Security Lab in Taiwan hosted by Auray
6) Asia and Pacific OTIC in PRC hosted by ZGC Academy
7) Kyrio O-RAN Test and Integration Lab in Louisville, CO hosted by CableLabs

11.3.7 O-RAN Global PlugFests

In order to keep up with the rapid pace of Open RAN development and continuing to support the ecosystem as their implementations mature and evolve, the O-RAN ALLIANCE also organizes Global PlugFests that allow the entire ecosystem of operators and vendors to come together and showcase the latest developments according to the most recent technical and test specifications. It started as annual events hosted by a handful of operators around the globe, to more recently a biannual event (one in the Spring and one in the Fall) and hosted by several operators and independent test houses/academic labs some of which are OTICs themselves. The principles of vendor-neutrality and openness are also upheld similar to OTICs, and are a great opportunity for the operator community to emphasize their priorities to the vendor ecosystem, and likewise for the vendor ecosystem to showcase their latest and greatest to the operator community.

11.4 Conclusion

In this chapter, we have outlined the main test and integration activities undertaken by the Open RAN ecosystem, and how the O-RAN ALLIANCE fosters this ecosystem through test specifications, a certification and badging program, OTICs, and global plugfests.

Bibliography

3GPP TS 33.511 (2020). *Security Assurance Specification (SCAS) for the Next Generation Node B (gNodeB) Network Product Class (Release 16)*.

O-RAN.TIFG.E2E-Test.0-v04.00 (2022). *O-RAN Test and Integration Focus Group End-to-End Test Specification*.

O-RAN.TIFG.Cert-Badge.0-v06.00 (2022). *O-RAN Test and Integration Focus Group Certification and Badging Processes and Procedures*.

O-RAN.TIFG.CGofOTIC.0-v04.00 (2022). *O-RAN Test and Integration Focus Group Criteria and Guidelines for Open Test and Integration Centers.*

O-RAN.WG1.O-RAN-Architecture-Description-v08.00. *O-RAN Architecture Description.*

O-RAN.WG2.A1TS-v03.00 (2022). *O-RAN Working Group 2 A1 Interface: Test Specification.*

O-RAN.WG3.E2TS.v02.00 (2022). *O-RAN Working Group 3 E2 Interface Test Specification.*

O-RAN.WG4.CONF.0-v07.00 (2022). *O-RAN Fronthaul Working Group Conformance Test Specification.*

O-RAN-WG4.CUS.0-v11.00 (2022). *O-RAN Fronthaul Working Group Control, User and Synchronization Plane Specification.*

O-RAN-WG4.MP.0-v11.00 (2022). *O-RAN Fronthaul Working Group Management Plane Specification.*

O-RAN.WG4.IOT.0-v09.00 (2022). *O-RAN Fronthaul Working Group Interoperability Test Specification.*

O-RAN.WG5.IOT.0-v06.00 (2022). *O-RAN Open F1/W1/E1/X2/Xn Interface Interoperability Test Specification.*

O-RAN.WG5.C.1-v09.00 (2022). *O-RAN Open F1/W1/E1/X2/Xn Interfaces Working Group NR C-Plane Profile.*

O-RAN.WG5.U.0-v05.00 (2022). *O-RAN Open F1/W1/E1/X2/Xn Interfaces Working Group NR U-Plane Profile.*

O-RAN.WG8.IOT.0-v06.00 (2022). *O-RAN Stack Reference Design Working Group Stack Interoperability Test Specification.*

O-RAN.WG9.XTRP-Test.0-v02.00 (2022). *O-RAN Open Xhaul Transport Working Group 9 Xhaul Transport Testing.* Germany: O-RAN ALLIANCE.

O-RAN.WG11.O-RAN-Security-Protocols-Specifications-v05.00 (2022). *O-RAN Working Group 11 Security Protocols Specification.* Germany: O-RAN ALLIANCE.

O-RAN.WG11.O-RAN-Security-Tests-Specifications-v03.00 (2022). *O-RAN Working Group 11 Security Tests Specification.* Germany: O-RAN ALLIANCE.

12

Other Open RAN Industry Organizations

Aditya Chopra[1], Manish Singh[2], Prabhakar Chitrapu[3], Luis Lopes[4], and Diane Rinaldo[5]

[1] Amazon Kuiper, Austin, TX, USA
[2] Dell, Round Rock, TX, USA
[3] SCF, St. Louis, MO, USA
[4] Qualcomm Incorporated, San Diego, CA, USA
[5] ORPC, Portland, ME, USA

12.1   Telecom Infra Project

The Telecom Infra Project (TIP) was formed by Meta in 2016 as an engineering-focused, collaborative methodology for building and deploying global telecom network infrastructure, with the goal of enabling a broad ecosystem of hardware and software vendors. It is now a global community of companies and organizations that are collaborating to develop, test, and deploy open, disaggregated, and standards-based solutions. TIP is accelerating the development of true multivendor, disaggregated Open-RAN solutions through an ecosystem and by leveraging the open interface specifications from O-RAN Alliance. Furthermore, TIP supports multiple Community Labs across the globe where different member companies collaborate to test and validate Open RAN solutions. TIP also regularly hosts Plugfest events to assess the specifications and ecosystem maturity.

To accelerate innovation and to break vendor lock-in from monolithic RAN solutions, TIP's Open-RAN program supports the development of disaggregated and interoperable radio access network (RAN) solutions based on mobile operator requirements. TIP is driving RAN disaggregation on two key vectors (Figure 12.1):

(1) **Hardware and software disaggregation** enables COTS platforms and silicon to be leveraged for building and deploying RAN infrastructure thereby bringing the benefits of competition and economies of scale to RAN equipment.
(2) **Functional disaggregation** across various RAN functional units such as the RU, DU, CU, RIC, xApps, and OAM/NMS with open interfaces, thus enabling multivendor components to readily interoperate which allows operators to source components from a diversified supply chain.

Open RAN: The Definitive Guide, First Edition. Edited by Ian C. Wong, Aditya Chopra, Sridhar Rajagopal, and Rittwik Jana.

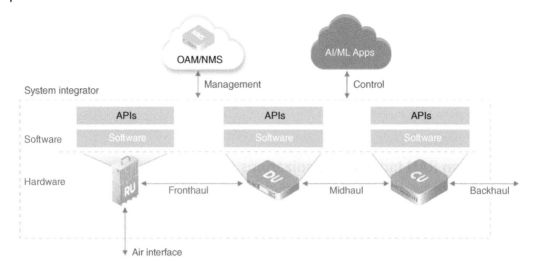

Figure 12.1 Hardware and software disaggregation in Open RAN. *Source:* TIP/https://telecominfraproject. com/openran/ last accessed 15 March 2023.

To enable such innovation and industry-wide collaboration, TIP has set up multiple Project Groups spanning across Access, Transport, and Core that collectively make up an end-to-end cellular network. All Open RAN-related work in TIP is conducted within the Open RAN Project Group, which has a mission to accelerate innovation and disaggregation in the RAN domain with multivendor interoperable products and solutions that are easy to integrate into the operator's network and are verified for different deployment scenarios. The Open RAN Project Group intends to enable abundant last-mile wireless connectivity by developing a vibrant ecosystem that is rapidly innovating at every layer, every component of the RAN technology stack, and continually improving the performance of RAN through innovation, automation, and competition.

12.1.1 Organizational Structure

Leadership for various activities in TIP's Open RAN Project Group is provided by mobile network operators (MNOs) acting as co-chairs of the Project Group (Open RAN Project Group 2020). RAN disaggregation efforts in TIP are logically structured as various subgroups under the Open RAN Project Group. There are component subgroups that are focused on developing requirements for various Open RAN components like RU, DU, RIC, etc. And there are segment subgroups that focus on system-level requirements and integration efforts (Figure 12.2).

The key component subgroups under the Open RAN Project Group are:

- **Radio Unit (RU):** This subgroup is defining the architecture and requirements for open and disaggregated RU, considering various deployment use cases including single-band, multiband, and massive-MIMO RUs. The subgroup also defines RU test plans and badging criteria for RUs.
- **Distributed Unit and Centralized Unit (DU and CU):** Historically, baseband units (BBUs) have been built with proprietary silicon or field-programmable gate arrays (FPGAs) and with closed interfaces. These have limited RAN innovation and limited deployment options for operators. The DU and CU subgroup is defining different deployment archetypes and associated hardware and software requirements for DU and CU functions. The subgroup efforts are targeted toward disaggregating BBU functions to open and disaggregated DU and CU functions

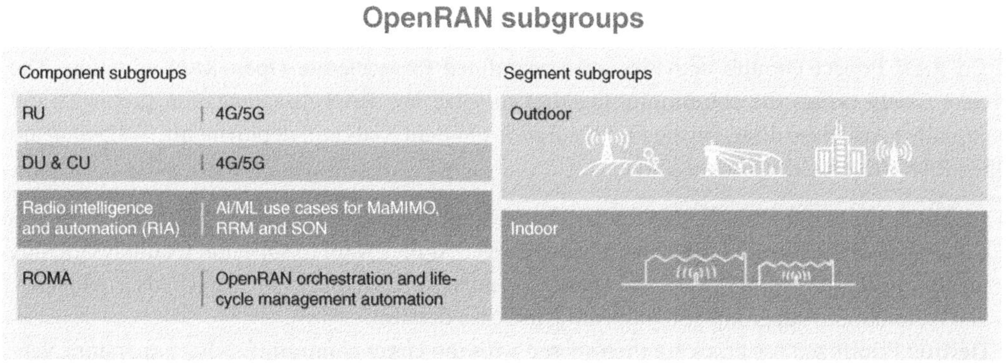

OpenRAN subgroups

Figure 12.2 TIP's Open RAN Project Group structure. *Source:* TIP/https://telecominfraproject.com/openran/ last accessed 15 March 2023.

that leverage merchant silicon, COTS platforms, and Cloud technologies. The subgroup also defines the DU and CU test plans and badging criteria.

- **RAN Intelligence and Automation (RIA):** Applying new technologies such as data science, artificial intelligence, and machine learning (AI/ML) to RAN will significantly increase performance, lower the cost of ownership via automation, and improve energy efficiency. RIA subgroup is working on several use cases that will automate network planning, configuration, and optimization in near-real-time to meet the growing and dynamic traffic demands. RIA subgroup has published a number of use cases for coverage and capacity optimization, interference mitigation, massive MIMO, and energy efficiency. This subgroup defines the RIC and xApps requirements for different use cases.
- **RAN Orchestration and Life Cycle Management Automation (ROMA):** Orchestration and life cycle management are key to accelerating the pace of innovation in RAN. ROMA subgroup is defining use cases, requirements, data models, and interfaces for RAN orchestration and automation. This will enable mobile operators to quickly test and deploy new features and functions in the production RAN. Further, the ROMA subgroup is working on automating Open RAN's life cycle management with Continuous Integration and Continuous Deployment (CI/CD) frameworks and tools.

In addition to the four component subgroups, the Open RAN Project Group has two segment subgroups:

- **Outdoor Macro:** Disaggregating RAN introduces integration and interoperability challenges. The outdoor macro subgroup develops reference architecture and system-level requirements for various outdoor segments including urban, periurban, and rural. The subgroup defines blueprints, integration test cases, and badging criteria for Open RAN solutions. To facilitate learning and accelerate Open RAN adoption, this subgroup also publishes reports from operator field trials and deployment playbooks.
- **Indoor Small Cells:** Disaggregation brings benefits to not only outdoor macro but also indoor small cells. Historically, small cells have been developed with closed interfaces and using proprietary silicon. This subgroup is defining multiple architectures for disaggregated Open RAN indoor small systems. The subgroup publishes hardware and software requirements for various modules and open interfaces. This subgroup also publishes test plans for indoor small cells and test reports from operator field trials to facilitate learning for the TIP community.

All the workstreams in both component and segment subgroups are mostly led by mobile operators and supported by vendors and even academia.

12.1.2 Core Activities

Open RAN Project Group's activities are streamlined to accelerate Open RAN adoption. The Project Group brings the community together to solve key RAN disaggregation problems and eventually takes those disaggregated RAN solutions to field trials and deployments. Following are Open RAN Project Group's core activities:

- **Ideate:** Most of the Open RAN Project Group activities start in the ideate phase where MNOs collaborate to identify and prioritize key use cases that they want to solve for. Examples include identifying massive MIMO, SON and energy optimization use cases in the RIA (RAN Intelligence and Automation) subgroup and publishing use case details.
- **Define:** Prioritized use cases are then shared with the larger community where operators, vendors, and even academia members collaborate to develop common requirements. Open RAN requirements are periodically published by the Open RAN Project Group. These published requirements form the basis to align the ecosystem – aligning vendor's roadmaps, integration resources in TIP community labs and also aligning the operator's trial.
- **Build:** In the build phase, TIP community members collaborate to define the blueprints, which include use cases, requirements, test plans as well as badge allocation criteria. Blueprint details are finalized and published in the Open RAN Project Group and are available to the entire community.
- **Test:** In the test phase, the community collaborates mostly in the TIP Community Labs. TIP Community Labs provide space and infrastructure for the community to come together and collaboratively integrate and test various Open RAN components coming from different vendors. TIP Community Labs make Open RAN integration efficient and ensure MNOs do not have to undertake basic integration efforts. Testing is done per the published test plans in the Project Group. Exit reports are published and TIP badges are allocated to solutions that successfully meet the criteria. TIP allocates different types of badges demonstrating the level of maturity of Open RAN solutions – more details on badge types are available on TIP Exchange (Telecom Infra Project 2022a).
- **Release:** Tested and badged products from different vendors are published on TIP Exchange. TIP Exchange is a marketplace where community vendors can publish their Open RAN products and MNOs can easily assess the availability and maturity of Open RAN solutions. Open RAN products validated in TIP Community Labs are listed on TIP Exchange with allocated badges, thereby clearly indicating the maturity of the products.
- **Deploy:** Adoption of Open RAN is a key objective of the Open RAN Project Group. After the solutions are validated and badged, the Project Group collaborates with MNO members to initiate trials and deployments. Taking a collaborative community approach, TIP enables MNOs to share trial results and deployment playbooks with the entire community. Examples include the published Open RAN Trials Vodafone Turkey Playbook (Telecom Infra Project 2022b).

While Open RAN Project Group publishes requirements, it does not publish interface specifications. Instead, the Project Group leverages interface specifications that are published by the O-RAN Alliance. TIP and the O-RAN Alliance have a Liaison Agreement that enables cross-referencing, information sharing, and validation activities.

12.2 Trials and Deployments

For RAN disaggregation to happen in the long term, it is critical that the technology sees the light of day in the real world. TIP's Open RAN Project Group works closely with mobile operators across the globe to initiate lab and field trials and to eventually start deploying Open RAN solutions at

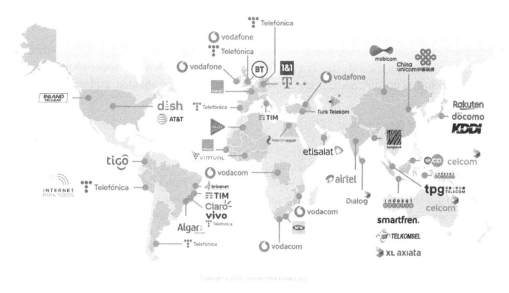

Figure 12.3 Open RAN trials and deployments. *Source:* TIP/https://telecominfraproject.com/
tip-openran-pg-accelerates-open-ran-commercialization/.

scale. Open RAN Project Group facilitates operators to holistically engage with the ecosystem, share their requirements, evaluate various solutions, and initiate trials. Most of the TIP Community Labs are hosted by mobile operators where initial integration and testing work is conducted before operators start their lab and field trials (Figure 12.3).

Every six months or so, TIP publishes a blog which includes the latest operator trials and deployments. Over the years, the number of operator trials have been growing significantly. The given map was published in TIP's blog at the TIP Insights 2021 event (Zani 2021).

While Open RAN has come a long way from its early days as a concept, with trials now underway across all major continents and various countries, we still have a long way ahead to make the disaggregated RAN a reality for mobile operators and societies at large. Abundant last-mile wireless connectivity is essential in making mobile internet a reality for everyone across the globe and Open RAN holds the promise to do just that. TIP's Open RAN Project Group is accelerating that journey.

12.3 Small Cell Forum

Small Cell Forum (SCF) is a global organization committed to the acceleration of the deployment of small cells in cellular networks. It strives to achieve this mission in a multipronged approach:

1. SCF conducts regular surveys and analyses of the small cell market and publishes the results as public documents. The results of such market analysis define and drive the work programs within SCF.
2. In order to accelerate small cell adoption, SCF identifies traditional and emerging deployers and operators of small cell networks and assists in their development needs. Such assistance can include the development of new standards, best practices, white papers, and engagement with appropriate regulatory bodies to address any regulatory hurdles.

3. SCF provides a platform for established and emerging vendors to participate in the market-place. This is done by developing core technical standards, and white papers on end-to-end architectures and management and operational solutions.

A testament to the success of the forum in meeting the given mission and objectives is the fact that SCF is seen widely in the industry as a trusted source of information, with over 100 published business and technical documents that are broadly downloaded, implemented, and shared.

12.3.1 A History of Openness at SCF

Since its inception and even before the terms of Open-RAN were coined, SCF has been pioneering the Open RAN concepts and associated technical and business solutions. For example, small cells were first introduced to the mobile communications industry as Femtocells, which were primarily for residential use and for addressing the poor indoor coverage problems in those situations. While these devices were really small cellular base stations in principle, there were fundamental differences that required a new set of open standards. For example, unlike conventional cellular base stations deployed in trusted environments and connected to the core networks using trusted transport, femtocells were deployed in untrusted (by the MNO) environments using untrusted customer transport to the core network. This required securitization of not only the femtocells themselves but also the transport by using IPSec techniques. The interface between the femtocells and core network was first pioneered in the SCF, which was later taken up and standardized by 3GPP. Unlike MNO-deployed cellular base stations, femtocells were also deployed in numbers which were larger by orders of magnitude, requiring SCF to publish open architectures which involved the use of femtocell aggregation gateways. Being customer premise devices, the operations and management of femtocells also required a different methodology than that of conventional cellular base stations. This led to a collaborative effort between SCF and BBF (Broadband Forum), producing new management data models such as TR196 using TR69 communication protocols.

The early recognition of the fact that the femtocells were a new business area offered a great opportunity for new, emerging, innovative, disruptive players, not only at the product level but also at the silicon and software levels. Therefore, SCF had to address the needs of the then emerging femtocell ecosystem. In particular, SCF saw the need to "disaggregate" HW and SW functions and created an open interface called FAPI (Femto API, which later became Functional API) between the physical layer HW (i.e. silicon solutions) and upper layer software. Starting with the 2G version of FAPI, SCF continued developing the FAPI specification for succeeding generations of the cellular industry, by addressing the targeted needs of the small cell industry. Specifically, the FAPI was extended to 3G, 4G (LTE), and 5G, spanning 3GPP releases up until Release-16. Functionally speaking, FAPI support extended to MIMO, Advanced Beam Management, IOT, etc. The FAPI solution was also extended to a networked model, called nFAPI, that allowed for a geographical separation between the physical layer hardware and upper layer software. SCF continues to work on developing the FAPI and nFAPI specifications and aligning these with the deployment architectures from 3GPP and O-RAN Alliance.

12.3.2 Alignment with the 3GPP and O-RAN Alliance Solutions

SCF recognizes that the increasingly diverse set of deployment environments for access networks comes with a diverse set of requirements and that one solution does not fit all situations. As such, SCF believes that a toolbox of disaggregated Open RAN solutions is needed by the emerging marketplace. It further follows that it is in the best interests of the entire mobile industry that these solutions are aligned as much as possible. SCF is striving to achieve this alignment with O-RAN Alliance by ensuring that its solutions are compatible with the overall Open RAN Architecture proposed by O-RAN Alliance, including the Management Solution framework. Indeed, it makes a lot of sense that the overall management and service orchestration solutions are common for all of the disaggregated Open RAN Solutions. The split option 6-based fronthaul specification published by SCF is one of the popular low-PHY RAN disaggregation solutions. The non-networked version of this specification, known as the FAPI interface, is even used within the O-RAN alliance as part of their baseband software specifications.

12.4 3rd Generation Partnership Project

The 3rd Generation Partnership Project (3GPP) is a collaboration between global telecommunications standards defining organizations to create technical specifications and standards for cellular mobile networks. As the name suggests, 3GPP was initially created with the goal of standardization of the third-generation (3G) mobile networks. However, their work has continued through the fourth-generation (4G) and long-term evolution (LTE) mobile standards, and on to the fifth-generation (5G) of mobile network standards dubbed as New Radio (NR).

The study and specification of RAN architecture and network interfaces are an integral part of standardization in 3GPP. 5G NR is certainly no exception – indeed, the discussion on NG (next generation)-RAN architecture started at the very beginning of the 5G standardization process in 3GPP. In (3GPP 2017), 3GPP captured the detailed findings of a one-year long study on architecture and interfaces for 5G NR. Similarly to the air interface design, NG-RAN is required to efficiently and effectively support a variety of use cases in various deployment scenarios utilizing a flexible range of frequencies (Chen et al. 2021). In particular, the following three high-level use cases form the foundation for 5G NR standardization:

- Enhanced Mobile Broadband (eMBB),
- Massive Machine Type Communications (mMTC), and
- Ultra-reliable and Low Latency Communications (URLLC).

These sets of use cases can be translated into a set of performance requirements in the form of reliability, availability, and latency, which drives the NG-RAN study (3GPP 2017).

The possibility of virtualizing NG-RAN drew strong interest at the start of the NG-RAN study, where a large number of split options were identified in (3GPP 2017), as illustrated in Figure 12.4. It is noted that the initial goal was to define a functional split between a Centralized Unit (CU) and a distributed unit (DU), in order to have a centrally and potentially virtualized CU (Chen et al. 2021).

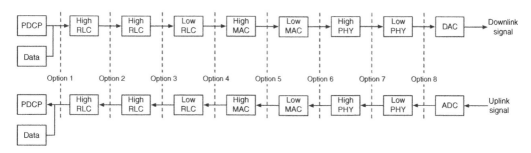

Figure 12.4 Illustration of potential function splits between central and distributed units.

As the standardization progressed, it became evident that these different split options offer the possibility of accommodating various deployment scenarios. At the same time, they also provide more options for commercial product offerings.

Standardization of a feature in 3GPP is, in general, very involved. With the participation of hundreds of delegates of different affiliations, the study of a feature usually results in many possible design options. Each option has its own pros and cons, in terms of standardization efforts, performance, complexity, commercial interests, etc. Convergence to a particular solution often requires extensive evaluation and analysis, and perhaps more importantly, constructive cooperation and flexibility from all participants. The resulting convergence, also known as *consensus* in 3GPP, is often a compromise from all parties involved, aiming for a common goal of delivering a quality specification for in-time and successful commercial deployments.

NG-RAN supports option 2 of the function split from the first 5G NR release (Release 15), as part of the so-called *higher-layer split*. Consequently, gNB-CU and gNB-DU are newly introduced for NG-RAN. Moreover, within a gNB-CU node, the split of the control plane (CP) entity and user plane (UP) entity is supported, resulting in gNB-CU-CP and gNB-CU-UP, respectively. These features enable NG-RAN for flexible deployments, especially for wide areas, where the CU node may be virtualized. However, due to the enormous time pressure in 5G NR standardization and the expected overwhelming amount of standardization efforts, *lower-layer split* options were not standardized in 3GPP.

Starting from Release 18, 5G NR is entering the second phase of standardization, the start of *5G-Advanced*. In (3GPP 2021b), the Release 18 package was approved, which contains numerous projects, as summarized and envisioned during the first RAN Release 18 workshop in (3GPP 2021a). The RAN Release 18 package provides a good balance in terms of:

- Balanced mobile broadband evolution vs. further vertical domain expansion,
- Balanced immediate vs. longer term commercial needs, and,
- Balanced device evolution vs. network evolution.

In particular, RAN Release 18 package contains several projects aiming for longer term commercial needs, e.g.:

- Study on Artificial Intelligence (AI)/Machine Learning (ML) for NR Air Interface
- Study on Evolution of NR Duplex Operation
- Study on low-power Wake-up Signal and Receiver for NR
- Study on enhancement for resiliency of gNB-CU

Note that there is also a project named "Artificial Intelligence (AI)/Machine Learning (ML) for NG-RAN," which is to start as a work item focusing on use cases identified during the Release 17 study phase, followed by a planned study item focusing on new use cases.

In particular, a RAN Release 18 project proposed in (NTT DOCOMO Inc 2021) focuses on the study and identification of failure scenarios associated with the gNB-CU-CP, based on the current architecture for the NG-RAN. This is motivated by the fact that for each split gNB in NG-RAN, since a single logical gNB-CU-CP is connected to multiple logical gNB-DUs and logical gNB-CU-UPs, it becomes necessary to investigate potential failure scenarios, which may lead to subsequent study of potential solutions.

As a global standardization body with hundreds of members from all regions, 3GPP has been open to new ideas, continuing to drive evolutions on various fronts. This is also witnessed by successful commercial deployments impacting the entire society. As we are entering the era of 5G-Advanced with 6G expected to arrive in the future, it is envisioned that 3GPP RAN architecture and interfaces will continue evolving.

12.5   Open RAN Policy Coalition

On 8 October 2012, the Chairman and Ranking Member – the top Republican and Democrat – of the prestigious House Permanent Select Committee on Intelligence dropped an explosive report that called into question the use of equipment from foreign adversaries in our telecommunications networks. Known as the "Huawei Report," it was an examination of the world's top two telecommunication network vendors: Huawei and ZTE, and their use of predatory practices to infiltrate global networks for nefarious activities.

Over the next several years, lawmakers continued to sound the alarm and slowly began a process of using regulatory actions that would make the purchase and installment of equipment from these companies next to impossible. Fast forward to 2017, and the global "Race to win 5G" is in full swing. The US government and its allies were concerned about using what could possibly be a tool for spying, manipulating, or exfiltrating data in the next generation of networks. Through a series of legislative and executive actions, the US government banned the use of Huawei in domestic networks and authorized $1.9 billion to "rip and replace" Chinese gear already deployed in rural areas of the country.

Soon the rip-and-replace program, combined with rapid 5G deployment in the U.S. collided with an existential question: if not them, then who? How would the industry continue to thrive and grow with only a small handful of vendors operating in this space? The RAN is one of the most expensive components needed to build out a modern wireless network. Every cell tower employs radio to intercept cellular communications, along with central and distributed units to determine how the signal will travel. With national carriers each deploying 70,000+ macro facilities, such a demand threatened to unravel an already stretched supply chain. This intense strain, with so few solutions to it, led lawmakers to act.

In Washington, as in many parts of the world, uncertainty breeds many ideas: some good, some not so good. The concept of Open RAN – a market-based approach – produced widespread bi-partisan support for revitalizing the telecommunications sector and spearheading a new innovative approach to networking. The concept seemed so simple. By standardizing the interfaces between the subcomponents – the radio, hardware, and software – you create a more modular structure. This in turn would drive competition at the subcomponent level which already has many more supplier options.

Open RAN is not a new concept but the concern over supply chain constraints heightened the awareness and interest. In January 2020, Senate powerhouse Mark Warner, Chair of the Senate Select Committee on Intelligence, introduced legislation incentivizing research and development of Open RAN. The USA Telecommunications Act was the first time many had a foray into Open RAN; the introduction spurred many conversations in and around the telecommunications industry in DC. A handful of companies contemplated the idea of forming an informal group of stakeholders to help inform the handful of policymakers interested in this new topic of the playing field. The Open RAN Policy Coalition represents a group of companies formed to promote policies that will advance the adoption of open and interoperable solutions.

What started out as 5–11 to 31 to now 60-plus members has been a complete whirlwind to a new burgeoning industry, and those on the ground floor. A handful of companies coming together to educate policymakers on 5G activity and concerns over the future of networking has morphed into a global association participating in countless meetings and seminars worldwide.

After eight months of debate, a relatively short amount for Congressional debate, the authorization of the USA Telecommunications Act was passed into law. The legislation authorized two new grant programs: The Public Wireless Fund to be administered by NTIA, and the Multilateral Fund, to serve as grants to more developing nations around the world to be administered by the State Department.

This cause for concern brought together political adversaries to draft one of the few pieces of bipartisan legislation in a caustic Congress: The United States Innovation and Competition Act (USICA). Championed by Senate Majority Leader Chuck Schumer, and Senate Republican Todd Young, this legislation laid out a pathway to funding these priorities.

At the time of this writing, the USICA package is in the nascent stages of being conferenced with the House companion legislation, America COMPETES (Creating Opportunities to Manufacturing Pre-eminence in Technology and Economic Strength). A conference consists of Senators and House Representatives coming together to iron out the differences between the two bills. It is believed that the conference will conclude in early summer producing a single bill that will get an up or down vote before moving to the President's desk for a final signature.

This conversation has quickly turned to a global audience. With deployments occurring around the globe, advocates, and supporters of Open RAN continue to educate policymakers around the world speaking to the benefits of vendor diversification and innovation in the ecosystem.

12.6 Conclusion

The Open RAN movement is large and ambitious in its scope of creating a cellular network architecture based on open specifications and standardized interfaces. It is hard to imagine a single standards defining organization tackling this challenge successfully. Fortunately, as this chapter has shown, there are multiple industry organizations sectioning off different pieces of the overall problem. These organizations also fully understand that they do not exist alone in a vacuum and liaison with each other to ensure compatibility in their approaches toward achieving the shared goals of an Open RAN.

Bibliography

3GPP (2017). *3rd Generation Partnership Project; Technical Specification Group Radio Access Network; Study on New Radio Access Technology: Radio Access Architecture and Interfaces (3GPP TR 38.801 v14.0.0). https://www.3gpp.org/ftp//Specs/archive/38_series/38.801/38801-e00.zip.*

3GPP (2021a). *Summary of RAN Rel-18 Workshop (3GPP TDoc#RWS-210659). https://www.3gpp.org/ftp/tsg_ran/TSG_RAN/TSGR_AHs/2021_06_RAN_Rel18_WS/Docs/RWS-210659.zip.*

3GPP (2021b). Status Report of *TSG RAN#94-e (3GPP TDoc#SP-211589). https://www.3gpp.org/ftp/tsg_sa/TSG_SA/TSGS_94E_Electronic_2021_12/Docs/SP-211589.zip.*

Chen, W., Gaal, P., Montojo, J. and Zisimopoulos, H. (2021). *Fundamentals of 5G Communications: Connectivity for Enhanced Mobile Broadband and Beyond.* McGraw Hill.

NTT DOCOMO Inc (2021). *Study on Enhancement for Resiliency of gNB-CU (3GPP TDoc#RP-213677). https://www.3gpp.org/ftp/TSG_RAN/TSG_RAN/TSGR_94e/Docs/RP-213677.zip.*

Open RAN Project Group (2020). *Playbook – OpenRAN Trials w/Vodafone Turkey.* https://cdn.brandfolder.io/D8DI15S7/as/c5tx5crn45cch6w3nrz39s/OpenRAN_VF_TK_Playbook_FINAL.pdf.

Telecom Infra Project (2022a). *Open RAN – Telecom Infra Project. Open RAN MoU Group Page.* https://telecominfraproject.com/openran.

Telecom Infra Project (2022b). *Exchange Portal – Telecom Infra Project.* TIP Badges and Ribbons. https://exchange.telecominfraproject.com/about-exchange/badges.

Zani, A. (2021). *TIP Insights 2021: Accelerating Commercial Realities.* https://telecominfraproject.com/tip-insights-2021-accelerating-commercial-realities.

Index

Open RAN: The Definitive Guide, First Edition. Edited by Ian C. Wong, Aditya Chopra, Sridhar Rajagopal, and Rittwik Jana.
© 2024 The Institute of Electrical and Electronics Engineers, Inc. Published 2024 by John Wiley & Sons, Inc.

Printed and bound by CPI Group (UK) Ltd, Croydon, CR0 4YY

16/04/2025

14658606-0003